T0205529

Intelligent Systems Reference Library

Volume 176

Series Editors

Janusz Kacprzyk, Polish Academy of Sciences, Warsaw, Poland

Lakhmi C. Jain, Faculty of Engineering and Information Technology, Centre for
Artificial Intelligence, University of Technology, Sydney, NSW, Australia;
KES International, Shoreham-by-Sea, UK;
Liverpool Hope University, Liverpool, UK

The aim of this series is to publish a Reference Library, including novel advances and developments in all aspects of Intelligent Systems in an easily accessible and well structured form. The series includes reference works, handbooks, compendia, textbooks, well-structured monographs, dictionaries, and encyclopedias. It contains well integrated knowledge and current information in the field of Intelligent Systems. The series covers the theory, applications, and design methods of Intelligent Systems. Virtually all disciplines such as engineering, computer science, avionics, business, e-commerce, environment, healthcare, physics and life science are included. The list of topics spans all the areas of modern intelligent systems such as: Ambient intelligence, Computational intelligence, Social intelligence, Computational neuroscience, Artificial life, Virtual society, Cognitive systems, DNA and immunity-based systems, e-Learning and teaching, Human-centred computing and Machine ethics, Intelligent control, Intelligent data analysis, Knowledge-based paradigms, Knowledge management, Intelligent agents, Intelligent decision making, Intelligent network security, Interactive entertainment, Learning paradigms, Recommender systems, Robotics and Mechatronics including human-machine teaming, Self-organizing and adaptive systems, Soft computing including Neural systems, Fuzzy systems, Evolutionary computing and the Fusion of these paradigms, Perception and Vision, Web intelligence and Multimedia.

** Indexing: The books of this series are submitted to ISI Web of Science, SCOPUS, DBLP and Springerlink.

More information about this series at http://www.springer.com/series/8578

Katarzyna Stapor

Introduction to Probabilistic and Statistical Methods with Examples in R

 Springer

Katarzyna Stapor
Faculty of Automatic Control,
Electronics and Computer Science
Silesian Technical University
Gliwice, Poland

ISSN 1868-4394 ISSN 1868-4408 (electronic)
Intelligent Systems Reference Library
ISBN 978-3-030-45801-0 ISBN 978-3-030-45799-0 (eBook)
https://doi.org/10.1007/978-3-030-45799-0

© Springer Nature Switzerland AG 2020
This work is subject to copyright. All rights are reserved by the Publisher, whether the whole or part of the material is concerned, specifically the rights of translation, reprinting, reuse of illustrations, recitation, broadcasting, reproduction on microfilms or in any other physical way, and transmission or information storage and retrieval, electronic adaptation, computer software, or by similar or dissimilar methodology now known or hereafter developed.
The use of general descriptive names, registered names, trademarks, service marks, etc. in this publication does not imply, even in the absence of a specific statement, that such names are exempt from the relevant protective laws and regulations and therefore free for general use.
The publisher, the authors and the editors are safe to assume that the advice and information in this book are believed to be true and accurate at the date of publication. Neither the publisher nor the authors or the editors give a warranty, express or implied, with respect to the material contained herein or for any errors or omissions that may have been made. The publisher remains neutral with regard to jurisdictional claims in published maps and institutional affiliations.

This Springer imprint is published by the registered company Springer Nature Switzerland AG
The registered company address is: Gewerbestrasse 11, 6330 Cham, Switzerland

Preface

This book *Introduction to Probabilistic and Statistical Methods with Examples in R* is the English and extended version of my previous textbook "*Lectures on statistical methods for IT students with examples in R*", published in Polish by Publishing House of Silesian University of Technology, Gliwice, Poland. This textbook is the foundation for one semester academic course "Probabilistic and statistical methods" which I have been giving at the Faculty for Automatic Control, Electronics and Computer Science of the Silesian University of Technology in Gliwice (Poland) for many years.

The course for which the materials were written is 50-50 Probability and Statistics, and the prerequisites for the course include mathematical analysis, and, in a few cases, linear algebra. The vision for this document is a more or less self-contained, essentially complete, correct introductory textbook. This book is intended primarily for students of different faculties of polytechnic schools but it will also be appropriate for university students of such faculties like biology, medicine, psychology, chemistry or economy.

This book can be subdivided into three basic parts or chapters:

1. Elements of the Probability Theory.
2. Descriptive and Inferential Statistics.
3. Correlation and Regression Analysis.

Chapter 1 includes the basics from probability theory: starting from the notion of probability space for random experiment it introduces then the concept of uni- and multivariate random variables together with different types of their distributions. Next, based on the short description of pseudo-random number generation, the examples of computer simulation of random events are given.

Chapter 2 starts with descriptive statistics which just explores the given data: describes and summarizes it with tables, charts and graphs. Then, the described sampling distributions pave the way to inferential statistics, which allows us to draw conclusions about the whole population from which we took the sample. We describe here the basics of statistical inference: estimation and hypothesis testing.

Chapter 3 presents one example of applied statistics: the elements of correlation and regression analysis. The first is used to quantify the association between two variables, while the second is a related technique to assess the relationship between an outcome variable (a response) and one or more explanatory variables (predictors).

At the end of the book, there are two appendices. The first, *"Permutations, combinations, variations"* gives the basic definitions from combinatorics. The second, *"Introduction to R"*, serves as an introduction to R language and it's uses, and provides an overview of how to implement some rudimentary statistical techniques and compute basic statistics.

The carefully selected examples throughout the book illustrate with real-world or simulated data, the methods of descriptive and inferential statistics. For some examples, their implementations in the R language are given. For the better readability, the relevant fragments with the R code are separated by the symbol "** R **".

Of course, there should be plenty of exercises for the student, with full solutions for some, and no solution for others. This will be covered in the exercise book which is planned in the near future.

Gliwice, Poland   Author
June 2019

Contents

Chapter 1
Elements of Probability Theory

1.1 Probability Space

Statisticians use the word *experiment* to describe any process that generates a set of data. A simple example of a *statistical experiment* is the tossing of a coin in which there are only two possible outcomes, heads or tails. Many board games require the rolling of an ordinary six-sided die. As in the case of a coin, before the die is rolled, we cannot predict the outcome. We cannot predict with certainty the number of accidents that occur at a given intersection of roads monthly or the class "defective" or "nondefective" of items going down from an assembly line, etc.

We are now ready to give a definition of a *statistical experiment* as the one in which there are a number of possible outcomes (greater than one), each possible outcome can be specified in advance, and we have no way of predicting which outcome will actually occur.

In this section, we will present a mathematical construct of probability theory, a *probability space* that is used as a model of a statistical experiment. We will also provide examples of probability spaces for various statistical experiments.

In probability theory a *probability space* is a construct consisting of three elements:

$$(\Omega, Z, P) \tag{1.1.1}$$

Ω a sample space,
Z a set of events (a σ—field of subsets of Ω),
P a probability measure on Z.

We now define the three elements of the probability space.

In probability theory, an *outcome* is a possible result of a statistical experiment. It is called a *simple* or *elementary event* and we denote it with a letter ω. Each possible outcome of a particular experiment is unique and only one outcome will occur on each trial of the experiment. All of the possible outcomes of an experiment form the elements of a *sample space* denoted with a letter Ω.

© Springer Nature Switzerland AG 2020
K. Stapor, *Introduction to Probabilistic and Statistical Methods with Examples in R*,
Intelligent Systems Reference Library 176,
https://doi.org/10.1007/978-3-030-45799-0_1

A set Z of *events* is defined as a family of subsets of a set Ω that satisfies the following conditions:

1. $\Omega \in Z$
2. $A \in Z \Rightarrow A^c = \Omega \backslash A \in Z$
3. $A_1, A_2, \ldots \in Z \Rightarrow \overset{\infty}{\underset{i=1}{\cup}} A_i \in Z$ $\qquad$ (1.1.2)

A set Z is, therefore, a σ-*algebra* or σ-*field* of subsets of Ω. The third condition states that a set Z is *closed* under the finite or countable union. Only the elements (subsets) of a set Z are called *events*. Thus, an *event* is a subset of points (i.e. elementary events) of a sample space Ω. The above stated requirements are necessary to assure a space (Ω, Z) of being a measurable space.

Before presenting the assignment of probabilities to events, we introduce some operations by which the new events are formed from the old ones.

Let J denotes a non-empty countable set of indicators.

Define a *union* of n events $A_i, i \in J$ denoted as:

$$\bigcup_{i \in J} A_i \qquad (1.1.3)$$

as a set consisting of those points of a sample space Ω which belong to at least one of the events $A_i, i \in J$.

An intersection of events $A_i, i \in J$ denoted as:

$$\bigcap_{i \in J} A_i \qquad (1.1.4)$$

is a set of points of a sample space Ω belonging to all of the events $A_i, i \in J$.

A *difference of events* A_1 and A_2 denoted as:

$$A_1 - A_2 \text{ or } A_1 \backslash A_2 \qquad (1.1.5)$$

is a set consisting of those points of sample space Ω that belong to A_1 and not belong to A_2.

We say that occurring of an event A_2 arises from a fact of occurring of an event A_1 which we denote as $A_1 \subset A_2$, if a set of points of a sample space Ω belonging to A_1 is included in a set A_2.

An event $A = \Omega$ is called a *sure event*, while an empty subset of a set Ω is called an *impossible* (*null*) *event* and denoted with a symbol $\emptyset$.

An *opposite event* A^c of an event A, also called a *complementary event* is defined as:

$$A^c = \Omega - A \qquad (1.1.6)$$

Two events A_1 and A_2 are said to be *mutually exclusive* or *disjoint* if the intersection of them is a null event:

$$A_1 \cap A_2 = \emptyset \tag{1.1.7}$$

The events A_i, $i \in J$ are said to be *pairwise mutually exclusive* if:

$$\underset{\substack{i,j \\ i \neq j}}{\forall} A_i \cap A_j = \emptyset \tag{1.1.8}$$

The events A_i, $i \in J$ are *collectively exhausted* if:

$$\bigcup_{i \in J} A_i = \Omega \tag{1.1.9}$$

that is, they collectively cover all points of a sample space Ω.

The events A_i, $i \in J$ are said to be *mutually exclusive and collectively exhausted* (MECE) *set of events* if the following two conditions are satisfied:

1. The events A_i are collectively exhausted:

$$\bigcup_{i \in J} A_i = \Omega \tag{1.1.10}$$

2. The events A_i are pairwise mutually exclusive.

A third element of a probability space is a *probability measure* P, which is a function that assigns a number $P(A)$ to each set A in a σ-field Z, provided that the following conditions (axioms) are satisfied:

1. $\forall A \in Z \;\; 0 \leq P(A) \leq 1$

2. $P(\Omega) = 1$ $\hspace{7cm}$ (1.1.11)

3. If $A_1, A_2, \ldots \in Z$ are any countable sequence of pairwise mutually exclusive events/sets in Z ($A_i \cap A_j = \emptyset$ for $i \neq j$) then:

$$P\left(\overset{\infty}{\underset{i=1}{\cup}} A_i \right) = \sum_{i \in 1}^{\infty} P(A_i) \tag{1.1.12}$$

A number $P(A)$ is called a *probability* of an event A. The first axiom states that a probability of an event is a non-negative real number from 0 to 1. According to the second axiom the probability that at least one of elementary events in the entire sample space will occur is 1 (the assumption of unit measure). The third axiom specifies the σ-additivity (the *countable additivity*) which means that a probability of an union of any countable sequence of pairwise mutually exclusive events in Z is equal to a sum of the probabilities of these events.

The above stated axioms are also known as the *axiomatic definition of probability* (proposed by the Russian mathematician A. N. Kolmogorov).

We will now consider the construction of probability spaces for the three types of sample spaces: finite, countable and uncountable ones.

Case 1. A Finite Sample Space
Consider a finite sample space:

$$\Omega = \{\omega_1, \omega_2, \ldots, \omega_n\} \tag{1.1.13}$$

Then the σ-algebra Z is defined by a power set of Z, that means $Z = 2^\Omega$. For any elementary event $\{\omega_i\} \in \Omega$, we can assign a probability $P(\{\omega_i\})$ such that

$$\sum_{i=1}^{n} P(\{\omega_i\}) = 1 \tag{1.1.14}$$

For any event $A \in Z$:

$$A = \{\omega_1, \omega_2, \ldots, \omega_m\} = \bigcup_{k=1}^{m} \{\omega_k\} \quad m \leq n \tag{1.1.15}$$

we can assign a probability $P(A)$ from the third axiom (1.1.12) constituting the definition of probability measure::

$$P(A) = \sum_{\omega_i \in A} P(\{\omega_i\}) \tag{1.1.16}$$

A probability $P(A)$ of event $A \in Z$ is a sum of the probabilities of elementary events $\{\omega_i\}$ constituting an event/subset A.

If we additionally assume that elementary *events are equally-likely*, we can assign equal probability $p = p_i = P(\{\omega_i\}) = 1/n$ to each elementary event $\{\omega_i\}$ (it is a uniform probability distribution). For any event $A \in Z$:

$$P(A) = P\left(\bigcup_{\omega_i \in A} \{\omega_i\}\right) = \sum_{i=1}^{m} P(\{\omega_i\})$$
$$= m \cdot \frac{1}{n} = \frac{m}{n} = \frac{|A|}{|\Omega|} \tag{1.1.17}$$

where symbol | . | denotes a number of elements of a set given in brackets. The formula:

$$P(A) = \frac{|A|}{|\Omega|} \qquad (1.1.18)$$

is the so-called *classical definition* of a probability, according to which finding a probability of an event A simply involves counting a number of elementary events constituting a set A (so-called "*favourable events*" to A) and divide it by a number of events forming a set Ω.

When n is large, counting by hand may not be feasible. *Combinatorial analysis* is simply a method of counting that can often be used to avoid writing down the entire list of favourable outcomes. The basic concepts of such counting a number of favourable events are given in Appendix A.

Case 2. A Countably Infinite Sample Space
Consider an infinite but countable sample space $\Omega = \{\omega_1, \omega_2, \ldots.\}$. We can construct a probability space as in the case 1. The σ-algebra Z of events is a power set of Ω. For any elementary event $\{\omega_i\} \in \Omega$, we can assign a probability $P(\{\omega_i\})$ such that

$$\sum_{i=1}^{\infty} P(\{\omega_i\}) = 1 \qquad (1.1.19)$$

For any event $A \in Z$ we assign probabilities:

$$P(A) = \sum_{\omega_i \in A} P(\{\omega_i\}) \qquad (1.1.20)$$

Case 3. An Uncountable Sample Space
Let a sample space is uncountable. Suppose $\Omega = R$. In such a case, the σ-algebra Z is a *Borel* σ-algebra (i.e. a smallest σ-algebra containing all open intervals, or roughly speaking, a class of sets obtained from intervals by countable number of such operations like union, intersection and complement). The elements of *Borel* σ-algebra are called *Borel sets*. So, the events are the Borel sets.

Let $f : R \rightarrow R$ is a non-negative integrable function such that,

$$\int_{\mathbb{R}} f(x)dx = 1 \qquad (1.1.21)$$

For any Borel set A,

$$P(A) = \int_{A} f(x)dx \qquad (1.1.22)$$

defines a probability on *Borel* σ-algebra.

As an example, suppose $\Omega = [a, b]$ and

$$f(x) = \begin{cases} \frac{1}{b-a} & \text{for } x \in [a, b] \\ 0 & \text{otherwise} \end{cases} \tag{1.1.23}$$

Then for $[a_1, b_1] \subseteq [a, b]$:

$$P([a_1, b_1]) = \frac{b_1 - a_1}{b - a} \tag{1.1.24}$$

This example paves the way to a geometrical definition of probability.

A geometrical probability of an event A is defined as a ratio of the Lebesgue'a measure (the area) of a set A to a measure of a sample space Ω:

$$P(A) = \frac{m(A)}{m(\Omega)} \tag{1.1.25}$$

The above formula determines the uniform probability distribution on a sample space Ω. The distribution is of continuous type, i.e. a probability of each one-point set is zero.

Below we present the examples of some specific probability spaces.

Example 1.1.1 Consider an experiment of tossing a fair coin. There are two possible outcomes (elementary events): H (head) and T (tail). The associated sample space is:

$$\Omega = \{H, T\} \tag{1.1.26}$$

A σ-*algebra* Z of events is a set of all subsets of Ω, it contains $2^{|\Omega|} = 2^2 = 4$ subsets:

$$Z = \{\emptyset, \{\Omega\}, \{H\}, \{T\}\} \tag{1.1.27}$$

Since the two elementary events are equiprobable, we get the following assignment of probabilities:

$$P(\{H\}) = P(\{T\}) = \frac{1}{2} \tag{1.1.28}$$

Example 1.1.2 Let us consider throwing a fair die. There are six possible outcomes: "one dot", "two dots", ... "six dots" resulting in a sample space:

$$\Omega = \{\{1\}, \ldots, \{6\}\} \tag{1.1.29}$$

A set Z of events contains of $2^6 = 64$ elements, these are an empty set, all possible 1-element subsets, 2-element subsets, ..., 5-element subsets, and only one 6-element subset, it is a sample space Ω.

Since the six elementary events are equiprobable, we get the following assignment of probabilities:

$$P(\{1\}) = \ldots = P(\{6\}) \frac{1}{6} \quad i = 1, 2, \ldots, 6 \tag{1.1.30}$$

Probabilities of other events from a set Z are calculated using the third axiom of the probability definition. Let's look for the interpretation of some of the events from a set Z. Events $A_1 = \{\{2\}, \{4\}, \{6\}\}$ and $A_2 = \{\{1\}, \{3\}, \{5\}\}$ can be described as "*the even/odd number of dots*". Events A_1 and A_2 are mutually exclusive events. The event $A_3 = \{\{1\}, \{2\}, \{3\}, \{4\}, \{5\}\}$ is a sure event: "*at least one of six dots*".

Example 1.1.3 Consider an experiment of tossing a fair coin until a head is obtained. We may have to toss a coin any number of times before a head is obtained. Thus, the possible outcomes are: *H, TH, TTH, TTTH*.... How many outcomes are there? The outcomes are countable but infinite in number. A countably infinite sample space is

$$\Omega = \{H, TH, TTH, \ldots\ldots\} \tag{1.1.31}$$

Let us denote

$$\omega_1 = H, \omega_2 = TH, \omega_3 = TTH \ldots\ldots \tag{1.1.32}$$

the elementary events $\omega_1, \omega_2, \omega_3, \ldots$ "Head in the 1st, 2nd, 3rd .... roll". We can assign the probabilities as:

$$P(\omega_n) = \frac{1}{2^n} \tag{1.1.33}$$

This is a special case of a so-called geometric distribution. Note that probabilities cannot be evenly distributed!

For any event A from a set Z, a P measure is calculated using the third axiom of probability. So, for example, for $A = \{\omega_1, \ldots, \omega_4\}$ we have:

$$P(A) = P(\{\omega_1\}) + P(\{\omega_2\}) + P(\{\omega_3\}) + P(\{\omega_4\})$$

$$= \sum_{n=1}^{4} \frac{1}{2^n} = \frac{15}{16} \tag{1.1.34}$$

Example 1.1.4 Consider an experiment of recording the time points $t_1, t_2, \ldots$ of the coming telephone calls to a certain person in an office in a period time from 8 a.m. to 16 p.m. The recorded time points $t_i \in [8, 16] = \Omega$ are the simple events. A set Z of events consists of the intervals included in a set $\Omega = [8, 16]$.

We will define a probability measure as the defined earlier geometrical probability. Consider any event from Z, e.g. $A = [14, 16] \subset [8, 16]$, meaning that somebody will call between 14 and 16. We have $P([14, 16]) = \frac{2}{8} = \frac{1}{4}$. Note, that for any single moment of t_i, a probability that someone will call is 0.

Example 1.1.5 Consider the popular sports lottery. The experiment relies on random drawing 6 out of 49 numbers. Before experiment, a player selects (crosses out) 6 numbers (from 49). Player's win depends on the level of agreement between the numbers selected and those drawn in experiment (the bigger the fitting the greater the win). We will build the probability space for this experiment and calculate the probability of the event $A = $ "3 *numbers fitted*".

The set of simple events Ω consists of all 6-element subsets (i.e. combinations) $\omega_i = \{\omega_{i_1}, \ldots, \omega_{i_6}\}$, $\omega_{i_k} \in \{1, 2, \ldots, 49\}$, $k = 1, 2, \ldots, 6$, which can be created from 49 numbers. The number of such combinations is:

$$\binom{49}{6} \tag{1.1.35}$$

The set Z of events is composed of all possible subsets of Ω. We assign equal probability to each elementary event ω_i:

$$p_i = \frac{1}{\binom{49}{6}} \tag{1.1.36}$$

The set A of favourable events consists of:

$$m = \binom{6}{3}\binom{43}{3} \tag{1.1.37}$$

elements, because from the 6 numbers that has been drawn you can create $\binom{6}{3}$ different triples. You can then add any three numbers from the remaining 43 to each of those triples. There are $\binom{43}{3}$ such triples and, therefore, there are:

$$\binom{6}{3}\binom{43}{3} \tag{1.1.38}$$

combinations favouring the event A. Thus, the probability of event A is:

$$P(A) = \frac{\binom{6}{3}\binom{43}{3}}{\binom{49}{6}} = 0.0177 \tag{1.1.39}$$

From the earlier stated axioms of probability (1.1.12), several basic results or properties can be established. We list only some of them.

1. $P(\emptyset) = 0$

2. $P(A^c) = P(\Omega) - P(A) = 1 - P(A)$

3. $P(A_1 \cup A_2) = P(A_1) + P(A_2) - P(A_1 \cap A_2)$ \hfill (1.1.40)

4. If $\forall_{i \neq j} A_i \cap A_j = \emptyset$ then $P\left(\cup_{i=1}^n A_i\right) = \sum_{i \in 1}^n P(A_i)$

For example, we will prove 3rd property, one of the additive rules.

Proof We have

$$A_2 = (A_1 \cap A_2) \cup (A_2 \backslash A_1) \tag{1.1.41}$$

from the 3rd axiom (1.1.12) we have:

$$P(A_2) = P(A_1 \cap A_2) + P(A_2 \backslash A_1) \tag{1.1.42}$$

Rewriting the above result, we obtain:

$$P(A_2 \backslash A_1) = P(A_2) - P(A_1 \cap A_2) \tag{1.1.43}$$

It may be helpful to write:

$$A_1 \cup A_2 = A_1 \cup (A_2 \backslash A_1) \tag{1.1.44}$$

as a union of disjoint sets. Then, based on the 3rd axiom of probability we have:

$$P(A_1 \cup A_2) = P(A_1) + P(A_2 \backslash A_1) \tag{1.1.45}$$

Substituting in the above formula for $P(A_2 \backslash A_1)$ the right side of expression (1.1.43), we obtain the final property:

$$P(A_1 \cup A_2) = P(A_1) + P(A_2 \backslash A_1)$$

$$= P(A_1) + P(A_2) - P(A_1 \cap A_2) \qquad (1.1.46)$$

1.2 Conditional Probability, Independence and Bernoulli Trials

When we are dealing with an event whose probability depends on an occurrence of another event, we are talking about conditional probability and we call such events *dependent*.

A probability of an event $A_i \in Z$ occurring when it is known that some event A_0 has occurred is called a *conditional probability* and is denoted by $P(A_i|A_0)$. A symbol "|" is usually read "the probability of A_i given A_0".

A *conditional probability of A_i given A_0* is defined by:

$$P(A_i|A_0) = \frac{P(A_i \cap A_0)}{P(A_0)} \text{provided } P(A_0) > 0 \qquad (1.2.1)$$

Events A_i and A_0 are called *independent* if a probability of one of them has no effect on a probability of occurrence of the other. This can be formally expressed as:

$$P(A_i|A_0) = P(A_i) \quad P(A_0) > 0 \qquad (1.2.2)$$

From the above two formulas, it follows that two events A_i and A_0 are independent if and only if:

$$P(A_i \cap A_0) = P(A_i) \cdot P(A_0) \qquad (1.2.3)$$

that is, a probability of an intersection of events is equal to a product of their probabilities. A finite set of events $A_1, \ldots, A_n$ is *pairwise independent* if the events in each pair are independent.

A finite set of events $A_1, \ldots, A_n$ is *mutually independent* if and only if for every $k \le n$ and for every k-element subset of events $\{A_i\}_{i=1}^{k}$:

$$P\left(\bigcap_{i=1}^{k} A_i\right) = \prod_{i=1}^{k} P(A_i) \qquad (1.2.4)$$

Example 1.2.1 Consider rolling a six-sided die twice. Let us determine the probability space of this experiment. The set Ω has 36 elementary events:

$$
\begin{array}{llllll}
11 & 21 & 31 & 41 & 51 & 61 \\
12 & 22 & 32 & 42 & 52 & 62 \\
13 & 23 & 33 & 43 & 53 & 63 \\
14 & 24 & 34 & 44 & 54 & 64 \\
15 & 25 & 35 & 45 & 55 & 65 \\
16 & 26 & 36 & 46 & 56 & 66
\end{array}
$$

The symbol "13", for example, means the event "one point" has occurred in a first roll and "three points" in a second. Each subset of the set Ω is an event, i.e., $Z = 2^{\Omega}$. The elementary events are equally likely, therefore, we can assign probabilities as:

$$
p_i = P(\omega_i) = \frac{1}{36} \quad i = 1, \dots, 36 \tag{1.2.5}
$$

Let's introduce the following denotation of events:

A_1 one point in a first roll,
A_2 two points in a second roll,
A_3 sum of points in two rolls is equal to 3.

The probability of the event $A_1 = \{11, 12, 13, 14, 15, 16\}$ is equal:

$$
P(A_1) = \frac{6}{36} = \frac{1}{6} \tag{1.2.6}
$$

Similarly, we calculate the probabilities of other events. Probability of the event $A_2 = \{12, 22, 32, 42, 52, 62\}$ is equal:

$$
P(A_2) = \frac{6}{36} = \frac{1}{6} \tag{1.2.7}
$$

and of the event $A_3 = \{12, 21\}$:

$$
P(A_3) = \frac{2}{36} \tag{1.2.8}
$$

Intuitively, it seems that the events A_1 and A_2 are independent because the result of the first roll has nothing common with the result obtained in the second roll. This fact follows from the equality:

$$
P(A_1 \cap A_2) = P(\{12\}) = \frac{1}{36} = \frac{1}{6} \cdot \frac{1}{6} = P(A_1) \cdot P(A_2) \tag{1.2.9}
$$

On the other hand, events A_1 and A_3 are dependent, because the occurrence of an event A_1 influences the probability of an event A_3:

$$P(A_3|A_1) = \frac{P(A_3 \cap A_1)}{P(A_1)} = \frac{\frac{1}{36}}{\frac{1}{6}} = \frac{1}{6} > P(A_3) = \frac{2}{36} \tag{1.2.10}$$

Formally, this dependence follows from the following fact:

$$P(A_1 \cap A_3) = P(\{12\}) = \frac{1}{36} \neq P(A_1) \cdot P(A_3) = \frac{1}{6} \tag{1.2.11}$$

Bernoulli Trials

A statistical experiment often consists of independent repeated trials, each with exactly two possible outcomes that may be labeled "success" (S) and "failure" (F). The probability p of success (or failure q) is the same each time the experiment is conducted. Such independent repeated trials of an experiment with exactly two possible outcomes are called *Bernoulli trials*. The probability of success and the probability of failure sum to unity (i.e. $p + q = 1$) since these are complementary events.

The most obvious example is flipping a coin in which obverse ("heads") conventionally denotes success and reverse ("tails") denotes failure. A fair coin has the probability of success $p = 0.5$ by definition.

One may be interested in probability of obtaining a certain number of successes, denote it k, in n Bernoulli trials. The following theorem enables computation of such probability.

Theorem 1.2.1 *The probability of obtaining k successes in the n Bernoulli trials is given by the formula:*

$$P_{n,k} = \binom{n}{k} p^k q^{n-k} \tag{1.2.12}$$

Proof Consider the following one possible result of n Bernoulli trials:

$$\underbrace{SSSS \ldots S}_{k \ times} \underbrace{FF \ldots F}_{n-k \ times} \tag{1.2.13}$$

consisting of k successes and $(n - k)$ failures. Due to the independence of individual events in the n trials, the probability of such result is:

$$p^k q^{n-k} \tag{1.2.14}$$

There are as many different orderings of k successes and $(n - k)$ failures as the permutations with repetitions of n elements, among which one repeats k times, and the other $(n - k)$ times (see Appendix A):

$$P_n^{k,n-k} = \frac{n!}{(n-k)!k!} = \binom{n}{k} \tag{1.2.15}$$

Therefore, the probability of obtaining k successes in n Bernoulli trials is:

$$P_{n,k} = \binom{n}{k} p^k q^{n-k} \qquad (1.2.16)$$

In the Bernoulli trials, the indicator for k_0 for which P_{n,k_0} is not smaller than the other probabilities is called the *most probable number of successes* in a series of n trials. This indicator satisfies the following inequalities:

$$(n+1)p - 1 \le k_0 \le (n+1)p \qquad (1.2.17)$$

If the numbers:

$$k_0^1 = (n+1)p - 1, \quad k_0^2 = (n+1)p \qquad (1.2.18)$$

are integers, then we have the two most likely numbers of successes. Otherwise, there is only one such number:

$$k_0 = [(n+1)p] \qquad (1.2.19)$$

where [.] means an integer part.

Example 1.2.2 The products manufactured by a factory are packed into packages of 10 items. The quality control states that 35% of packages on average contain damaged items. A customer has purchased 5 packages. What is the probability of an event that 3 of purchased packages will contain items without damage? What is the most likely number of undamaged packages?

The purchase of a single package can be treated as a Bernoulli trial. The success is a purchase of a package without defective items inside, and its probability is $p = 1 - 0.35 = 0.65$. Therefore, the probability of purchasing exactly 3 packages without damaged items is:

$$P_{5,3} = \binom{5}{3}(0.65)^3(0.35)^2 = 0.3364 \qquad (1.2.20)$$

Because:

$$k_0 = [(5+1) \cdot 0.65] = [3.9] = 3 \qquad (1.2.21)$$

so, the most likely number of undamaged packages is 3.

1.3   Theorem of Total Probability, Bayes Theorem and Applications

Theorem 1.3.1 of Total Probability *Let (Ω, Z, P) be a given probability space and $A_1, \ldots\ldots, A_n \in Z$ are the MECE events. For any event $A \in Z$, its probability can be computed using the following formula:*

$$P(A) = \sum_{i=1}^{n} P(A|A_i)P(A_i) \tag{1.3.1}$$

Proof An event A can be represented in terms of MECE events in the following way:

$$A = A \cap \Omega = A \cap \left(\bigcup_{i=1}^{n} A_i \right) = \bigcup_{i=1}^{n} A \cap A_i \tag{1.3.2}$$

Since the events $A \cap A_i$ ($i = 1, \ldots, n$) are mutually exclusive (which follows from MECE of $A_1, \ldots\ldots, A_n \in Z$), based on the 3rd axiom of probability, we can write the probability of event A as:

$$P(A) = P\left(\bigcup_{i=1}^{n} A \cap A_i \right) = \sum_{i=1}^{n} P(A \cap A_i)$$
$$= \sum_{i=1}^{n} P(A|A_i)P(A_i) \tag{1.3.3}$$

The probabilities $P(A_i)$ are called a priori *probabilities*, i.e. they are given in advance. The probability $P(A)$ defined by Formula (1.3.1) is called the *total probability*. The above theorem can be given the following interpretation. Having a given set of mutually exclusive events A_i in probability space (Ω, Z, P), whose union is a sure event, it is possible to calculate the probability of any other event $A \in Z$, based on a priori probabilities $P(A_i)$ of those events and the conditional probabilities $P(A|A_i)$ for $i = 1, \ldots, n$.

Example 1.3.1 We have three containers A_1, A_2, A_3 with some products among which there are also defective ones. There are 1% defective products in container A_1, 3% defective products in containers A_2 and 10% in container A_3. All the containers have the same chance to be selected. We randomly select a container, and, then, the product from this container. We're interested in a probability that the selected product is defective.

Let us denote by A_k ($k = 1, 2, 3$) an event that the kth container is selected. These events are the MECE events. A priori probabilities of these events are the same:

$$P(A_k) = \frac{1}{3} \quad k = 1, 2, 3 \tag{1.3.4}$$

The conditional probabilities follow from the given assumptions about defectivity:

$$P(A|A_1) = 0.01$$
$$P(A|A_2) = 0.03$$
$$P(A|A_3) = 0.1 \tag{1.3.5}$$

Using the theorem of total probability, we compute the probability of event A that a selected product is defective:

$$P(A) = \frac{1}{3}(0.01 + 0.03 + 0.1) \approx 0.05 \tag{1.3.6}$$

The Bayes' Rule

Continuing the Example 1.3.1, suppose now that the product was randomly selected (we don't know from which container), and it is defective. Instead of asking about $P(A)$, we are now asking about $P(A_1|A)$, the probability that this product was selected from the container A_1. Questions of this type can be answered using the following theorem, called the *Bayes rule*.

Let (Ω, Z, P) be a given probability space in which some events $A_1, \ldots, A_n \in Z$, named here as *hypotheses*, are defined, that are MECE events such that $P(A_i) \neq 0$ for $i = 1, \ldots, k$. Then, for any event $A \in Z$ such that $P(A) \neq 0$, called here an *evidence*, the *Bayes rule* derives *a posteriory probability* $P(A_i|A)$, the probability of A_i given A, i.e. after event A is observed, in the following way:

$$P(A_i|A) = \frac{P(A|A_i)P(A_i)}{\sum_{j=1}^{n} P(A|A_j)P(A_j)} \quad i = 1, \ldots, n \tag{1.3.7}$$

This is what we want to know: the probability of a hypothesis A_i given the observed evidence A.

Proof

From the definition of conditional probability we have:

$$P(A_i|A) = \frac{P(A \cap A_i)}{P(A)} \tag{1.3.8}$$

$$P(A|A_i) = \frac{P(A \cap A_i)}{P(A_i)}$$

We can substitute the nominator of the first expression with the right side of $P(A \cap A_i) = P(A|A_i)P(A_i)$ (this follows from the second expression) and the denominator using theorem of total probability which results in:

$$P(A_i|A) = \frac{P(A \cap A_i)}{P(A)} = \frac{P(A|A_i)P(A_i)}{\sum_{j=1}^{n} P(A|A_j)P(A_j)} \tag{1.3.9}$$

Continuing the Example 1.3.1, using the Bayes' rule, we can now compute a posteriori probability $P(A_1|A)$ that the randomly selected, defective product comes from the first container A_1:

$$P(A_1|A) = \frac{0.01 \cdot 1/3}{1/3(0.01 + 0.03 + 0.1)} = \frac{1}{14} \qquad (1.3.10)$$

The next example illustrates the nature of *Bayesian inference*: how the successive application of Bayes rule enables to update the probability for a hypothesis as more evidence becomes available.

Example 1.3.2 A patient comes to a doctor and reports some complaints. Based on symptoms doctor suspects certain disease H (an initial diagnosis). The frequency of disease H in general population is about 3%. This is a priori probability of an event $H+$ that a patient is ill with this disease: $p(H+) = 0.03$. A doctor orders the appropriate blood test T which appears to be positive (event $T+$). The following two facts are also known: *sensitivity* of the T test, i.e. the probability of detecting a disease in a group of patients with this disease is $p(T+ \mid H+) = 0.90$. *The specificity* of the T test, i.e. the probability of negative result of the test T is $p(T- \mid H-) = 0.94$.

The probability of false-positive results is, therefore

$$p(T + |H-) = 1 - P(T - |H-) = 1 - 0.94 = 0.06 \qquad (1.3.11)$$

Using the Bayes' rule, we can now compute how the positive result of the test T (i.e. an event $T+$) will modify a priori probability $p(H+)$, the level of our belief in an initial diagnosis:

$$
\begin{aligned}
p(H + |T+) &= \frac{p(T + |H+)p(H+)}{p(T + |H+)p(H+) + p(T + |H-)p(H-)} \\
&= \frac{0.90 \times 0.03}{0.90 \times 0.03 + 0.06 \times 0.97} = 0.3146 \approx 31.5\%
\end{aligned}
\qquad (1.3.12)
$$

After a positive test result has been occurred, the level of our belief in an initial diagnosis increases significantly: a small initial a priori probability 3% increases 10 times, i.e. up to 31.5%, but it is still more likely that the patient is not ill with a disease H. Do not rely on the results of a single test!

Therefore, the doctor orders another test U, now more specific to a disease H. The sensitivity test of the test U is $p(U + |H+) = 0.95$ and the specificity $p(U - |H-) = 0.97$.

The probability of false positive results is, therefore:

$$p(U + |H-) = 0.03 \qquad (1.3.13)$$

We now take $p(H+) = 0.3146$ as a priori probability, i.e. the previously obtained final probability. Then, the positive result of the test U (i.e. an event $U+$), increases the level of our belief in an initial diagnosis up to 94%:

$$
\begin{aligned}
p(H + |U+) &= \frac{p(U + |H+)p(H+)}{p(U + |H+)p(H+) + p(U + |H-)p(H-)} \\
&= \frac{0.95 \times 0.3146}{0.95 \times 0.3146 + 0.03 \times 0.6854} = 0.9356 \approx 93.6\% \quad (1.3.14)
\end{aligned}
$$

The strong increase of a posteriori probability illustrates the importance of a positive result in the second test.

1.4 Random Variable and Probability Distribution

1.4.1 A Concept of Random Variable

Statistics is concerned with making inferences about population characteristics. To be able to make such inferences, first, the statistical experiments are conducted. Their results are subject to chance. A good example of such statistical experiment is the testing of a number of defects in the products manufactured by some factory by a quality control. Suppose that 3 different types of defects can occur in products which can be detected by special tests used by quality control specialists. The sample space giving a detailed description of each possible outcome in this experiment is: $\Omega = \{0, 1, 2, 3\}$. The outcomes are the numbers of defects in a product randomly selected into a sample for quality control (statisticians use *sampling plans* to accept or reject batches of products). These numbers are, of course, random quantities determined by the outcome of the experiment. They may be viewed as the values assumed by the so-called *random variable* X, here being the number of defects in a tested product.

Formally, for a given probability space (Ω, Z, P) of a given statistical experiment, a *random variable* X is defined as a function:

$$
X : \Omega \to R \tag{1.4.1.1}
$$

that assigns to each elementary event $\omega \in \Omega$ a real number $X(\omega) \in R$ in such a way that for each $r \in R$ a set:

$$
\{\omega \in \Omega : X(\omega) < r\} \tag{1.4.1.2}
$$

is an event belonging to a set Z. The condition (1.4.1.2) is called the *measurability of a function X with respect to Z*. We will assume that all random variables presented in this textbook are measurable functions.

Thus, a *random variable* is a function that associates a real number with each elementary event in a sample space Ω.

We shall use a capital letter, say X, to denote a random variable and the corresponding small letter, x, for one of its values.

In the following examples which illustrate a notion of random variable, we assume that a probability space (Ω, Z, P) has been defined for a given statistical experiment.

Example 1.4.1.1

1. Consider rolling a six-sided die as a statistical experiment. We can define the random variable X that assigns to each elementary event $\omega_i \in \Omega$ a number of dots on a die's wall that has been occurred:

$$X(\omega_i) = i \quad i = 1, \ldots, 6 \tag{1.4.1.3}$$

2. Consider the mentioned statistical experiment of testing defects in the products manufactured by some factory. We can define a random variable X that assigns to each randomly selected product (elementary event ω) the number of defects in it:

$$X(\omega) = i \quad i \in \{0, 1, 2, 3\} \tag{1.4.1.4}$$

3. Consider a randomly selected inhabitant of a certain district that arrives to a tram stop without looking at the timetable and the trams run regularly every 10 min. We can define a random variable X that assigns to each random inhabitant's arriving (i.e. elementary event ω) the time of its waiting for a tram:

$$X(\omega) = t \in [0, 10] \tag{1.4.1.5}$$

A random variable concept permits us to replace a sample space Ω of arbitrary elements by a new sample space R having only real numbers as its elements. Furthermore, most problems in science and engineering deal with quantitative measures. Consequently, sample spaces associated with many random experiments of interest are already themselves sets of real numbers. A real-number assignment procedure is, thus, a natural unifying agent.

A random variable X is called a *discrete random variable* if it is defined over a sample space having a finite or a countably infinite number of sample points. In this case, random variable X takes on discrete values, and it is possible to enumerate all the values it may assume.

In the case of a sample space having an uncountably infinite number of sample points, the associated random variable is called a *continuous random variable*, with its values distributed over one or more continuous intervals on the real line. We make this distinction because they require different probability assignment considerations.

The behaviour of a random variable is characterized by its *probability distribution*, that is, by the way probabilities are distributed over the values it assumes. A *cumulative distribution function* and a *probability mass function* are two ways to characterize this distribution for a discrete random variable. They are equivalent in the sense that the knowledge of either one completely specifies the random variable.

The corresponding functions for a continuous random variable are the *cumulative distribution function*, defined in the same way as in the case of a discrete random variable, and the *probability density function*. The definitions of these functions will follow.

Given a statistical experiment with its associated random variable X and given a real number x, let us consider the probability of the event $\{\omega : X(\omega) < x\}$, or, simply, $P(X < x)$. This probability is clearly dependent on the assigned value x. The function $F_X : R \rightarrow [0, 1]$ defined by the formula:

$$F_X(x) = P(\{\omega : X(\omega) < x\}) = P(X < x) \quad x \in R \qquad (1.4.1.6)$$

is defined as the *cumulative distribution function* (cdf) of X. The cdf of a random variable thus accumulates probability as x increases. The subscript X identifies the random variable. This subscript is sometimes omitted when there is no risk of confusion. Thus, the value of a probability distribution at point x is the probability that a random variable X will take a value lower than the value x.

We give below some of the important properties possessed by cdf.

- It is a nonnegative, continuous-to-the-left, and nondecreasing function of the real variable x. Moreover, we have:

$$F(+\infty) = \lim_{x \to \infty} F(x) = 1 \quad F(-\infty) = \lim_{x \to -\infty} F(x) = 0 \qquad (1.4.1.7)$$

- It exists for discrete and continuous random variables and has values between 0 and 1.
- If a and b are two real numbers such that $a < b$, then

$$P(a < X < b) = F_X(b) - F_X(a) \qquad (1.4.1.8)$$

1.4.2 A Discrete Random Variable

Let X be a discrete random variable that assumes at most a countably infinite number of values $x_1, x_2, \ldots$ with nonzero probabilities. If we denote:

$$P(X = x_i) = p_i \quad i = 1, 2, \ldots \qquad (1.4.2.1)$$

then, clearly,

Table 1.1 A probability function of a discrete random variable

x_1	x_2	...	x_i	...
p_1	p_2	...	p_i	...

$$\begin{cases} 0 \le p_i \le 1 & \text{for all } i \\ \sum_i p_i = 1 \end{cases} \tag{1.4.2.2}$$

A function:

$$p_X(x) = P(X = x) \tag{1.4.2.3}$$

is defined as a *probability mass function* (pmf) of X, or, sometimes, *probability function*. Again, a subscript X is used to identify an associated random variable. A probability function can be given in a form of a formula or the following Table 1.1.

We also observe that, like $F_X(x)$, a specification of $p_X(x)$ completely characterizes random variable X. Furthermore, these two functions are simply related by:

$$p_X(x_i) = F_X(x_i) - F_X(x_{i-1}) \tag{1.4.2.4}$$

$$F_X(x) = \sum_{i=1}^{i:x_i \le x} p_X(x_i) \tag{1.4.2.5}$$

The upper limit for the sum in Eq. (1.4.2.5) means that the sum is taken over all i satisfying $x_i \le x$. Hence, we see that a cdf and pmf of a discrete random variable contain the same information; each one is recoverable from the other.

Example 1.4.2.1 Consider the simple game involving rolling a six-sided die. If the number of occurring dots is divided by 3, the player wins 100 \$, otherwise he/she loses 100 \$. Let a discrete random variable X associates each elementary event ω_i from sample the space $\Omega = \{\omega_1, \omega_2, \omega_3, \omega_4, \omega_5, \omega_6\}$ with a real number being his/her wins or loss as shown in the following Table 1.2.

The assignment defined in the table above defines the random variable X. Let's define now the distribution of this random variable X. Based on the 3rd axiom of probability, we first calculate the following probabilities:

Table 1.2 Specification of the random variable X from Example 1.4.2.1

	ω_1	ω_2	ω_3	ω_4	ω_5	ω_6
$x_i = X(\omega_i)$	-100	-100	$+100$	-100	-100	$+100$

Table 1.3 The pmf of random variable X from Example 1.4.2.1

x	-100	$+100$
$P(X = x)$	$\frac{2}{3}$	$\frac{1}{3}$

$$P(X = +100) = P(\{\omega_3, \omega_6\}) = \frac{1}{6} + \frac{1}{6} = \frac{1}{3} \qquad (1.4.2.6)$$

$$P(X = -100) = P(\{\omega_1, \omega_2, \omega_4, \omega_5\}) = 4 \cdot \frac{1}{6} = \frac{2}{3} \qquad (1.4.2.7)$$

The probability mass function of the random variable X is presented in the Table 1.3 and its plot in the Fig. 1.1.

The cumulative distribution function of our random variable X is:

$$F(x) = \begin{cases} 0 = P(\emptyset) & x \leq -100 \\ \frac{2}{3} = P(\emptyset \cup \{1, 2, 4, 5\}) & -100 < x \leq +100 \\ 1 = P(\emptyset \cup \{1, 2, 4, 5\} \cup \{3, 6\}) & +100 < x \end{cases} \qquad (1.4.2.8)$$

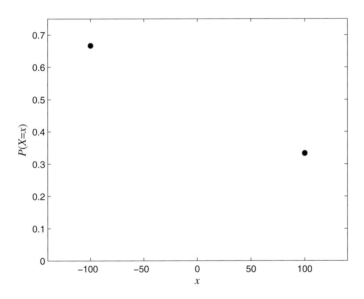

Fig. 1.1 Plot of pmf of random variable X from Example 1.4.2.1

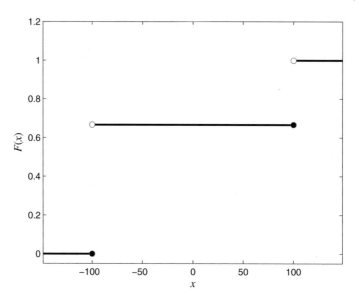

Fig. 1.2 Plot of cdf of random variable X from Example 1.4.2.1

The Fig. 1.2 shows a 'staircase' plot of the calculated cdf of X which is typical for a discrete random variable.

1.4.3   A Continuous Random Variable

For a continuous random variable X, its cdf, $F_X(x)$, is a continuous function of x and a derivative

$$f_X(x) = \frac{dF_X(x)}{dx} \tag{1.4.3.1}$$

exists for all x. A function $f_X(x)$ is called a *probability density function* (pdf), or simply a *density function* of X. Since $F_X(x)$ is monotone nondecreasing, we clearly have $f_X(x) \geq 0$ for all x. Additional properties of $f_X(x)$ can be derived easily from Eq. (1.4.3.1). These include:

$$\underset{x \in R}{\forall} \quad F_X(x) = \int_{-\infty}^{x} f_X(t)dt \tag{1.4.3.2}$$

$$\int_{-\infty}^{+\infty} f_X(x)dx = 1 \tag{1.4.3.3}$$

$$P(a \le X < b) = P(a < X \le b) = P(a < X < b)$$

$$= P(a \le X \le b) = \int_a^b f_X(x)dx = F_X(b) - F_X(a) \quad (1.4.3.4)$$

Example 1.4.3.1 Let us consider random sampling of a number from the range $[0,1]$, constituting the sample space Ω. Let us specify the random variable X as follows:

$$X(\omega) = \omega \quad \omega \in [0, 1] \quad (1.4.3.5)$$

i.e. the value of variable X for the particular number selected from the interval $[0, 1]$ is equal to the number that has been selected.

The distribution of this random variable is described by the following probability density function:

$$f_X(x) = \begin{cases} 1 \ x \in [0, 1] \\ 0 \ x \in R\backslash[0, 1] \end{cases} \quad (1.4.3.6)$$

The above defined random variable is a model of random "uniform" selection of numbers from the range $[0, 1]$. The cumulative distribution function of this random variable has the form:

$$F_X(x) = \begin{cases} 0 = \int_{-\infty}^{x} 0 \cdot ds & x \le 0 \\ x = \int_{-\infty}^{0} 0 \cdot ds + \int_0^x 1 \cdot ds = s \Big|_0^x & 0 < x \le 1 \\ 1 = \int_{-\infty}^{0} 0 \cdot ds + \int_0^1 1 \cdot ds + \int_1^x 0 \cdot ds \ x > 1 \end{cases} \quad (1.4.3.7)$$

Figure 1.3 shows the plots of probability density and cumulative distribution functions of the defined random variable X. The plot of cdf (dashed line) is typical for continuous random variables. It has no jumps or discontinuities as in the case of a discrete random variable. A continuous random variable assumes a none numerable number of values over a real line. Hence, a probability of a continuous random variable assuming any particular value is zero and, therefore, no discrete jumps are possible for its cdf. A probability of having a value in a given interval is found by using the Eq. (1.4.3.4). A pdf of a continuous random variable plays exactly the same role as a pmf of a discrete random variable. A function $f_X(x)$ can be interpreted as a mass density (mass per unit length).

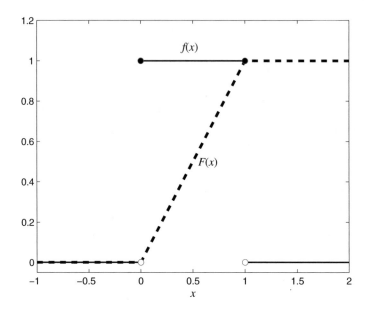

Fig. 1.3 Plot of pdf (solid line) and cdf (dashed line) of random variable X (Example 1.4.3.1)

1.4.4 *Parameters of Random Variable*

While a probability distribution $(F_X(x), f_X(x)$, or $p_X(x))$ contains a complete description of a random variable X, it is often of interest to seek for a set of simple numbers that gives a random variable some of its dominant features. These numbers include *moments* of various orders associated with a variable X. They are often called *measures* or *parameters* of a random variable and describe the most important numerical characteristics of a distribution of X.

An expectation (or an *expected value*) of a discrete random variable X, denoted by $E(X)$, is defined by the formula:

$$E(X) = \sum_i x_i \cdot p_X(x_i) \tag{1.4.4.1}$$

When the range of i extends from 1 to infinity, the sum in Eq. (1.4.4.1) exists if it converges absolutely.

An *expectation* of a continuous random variable X is defined by the formula:

$$E(X) = \int_{-\infty}^{+\infty} x \cdot f_X(x)dx \tag{1.4.4.2}$$

if the improper integral is absolutely convergent. The symbol $E(.)$ is regarded here and in the sequel as the *expectation operator*.

Let us note two basic properties associated with the expectation operator. For any real constants a, b, we have

$$E(b) = b \tag{1.4.4.3}$$

$$E(aX + b) = aE(X) + b \tag{1.4.4.4}$$

The above relations follow directly from the definition of $E(.)$.

The "moments" of a random variable (or of its distribution) are the expected values of powers of the random variable. *The k-th moment of a random variable X is defined as*:

$$m_k = E(X^k) = \begin{cases} \sum_i x_i^n \cdot p_X(x_i) & X \; discrete \\ \int_{-\infty}^{+\infty} x^n \cdot f_X(x)dx & X \; continuous \end{cases} \tag{1.4.4.5}$$

In particular, the first moment $m_1 = E(X)$ of X is its expected value. It is a measure of a "center" or "location" of a distribution. Using a mass analogy for a probability distribution, the first moment of random variable X may be regarded as a center of mass of its distribution. It is, thus, the average value of random variable X and certainly reveals one of the most important characteristics of its distribution. The first moment of X is synonymously called the *mean, expectation*, or *average value* of X.

The central moments of a random variable X are the moments of X with respect to its mean m_1. Formally, a *kth central moment* of a random variable X is defined as:

$$\mu_k = E(X - E(X))^k = \begin{cases} \sum_i (x_i - m_1)^n \cdot p_X(x_i) & X \; discrete \\ \int_{-\infty}^{+\infty} (x_i - m_1)^n \cdot f_X(x)dx & X \; continuous \end{cases} \tag{1.4.4.6}$$

In particular, the second central moment of a random variable X is its variance:

$$\mu_2 = E(X - E(X))^2 = D^2(X) \tag{1.4.4.7}$$

For a discrete random variable, a variance is reduced to a formula:

$$D^2(X) = \sum_i (x_i - E(X))^2 p_X(x_i) \tag{1.4.4.8}$$

while for a continuous random variable to:

$$D^2(X) = \int\limits_{-\infty}^{+\infty} (x - E(X))^2 f_X(x)dx \qquad (1.4.4.9)$$

A *standard deviation* of a random variable X is a (nonnegative) square root of variance:

$$D(X) = \sqrt{D^2(X)} \qquad (1.4.4.10)$$

The variance and standard deviation are the measures of spread, or, dispersion of a distribution. A standard deviation is measured in the same units as X, while a variance is in X-units squared. Large values of $D^2(X)$ imply a large spread in a distribution of X about its mean. Conversely, small values imply a sharp concentration of a mass of distribution in a neighborhood of a mean.

For a variance to exist, it must be assumed that $E(X^2) < \infty$. We note two properties of a variance of random variable X. They are:

$$D^2(b) = 0 \qquad (1.4.4.11)$$

$$D^2(aX + b) = a^2 D^2(X) \qquad (1.4.4.12)$$

for any real constants a, b.

The *quantile function* $Q(p)$ (where $0 < p < 1$) associated with a probability distribution of a random variable X, returns the value $x(p)$ of a random variable X such that the probability of variable X being less than that value $x(p)$ equals a given probability p. In the case of a continuous random variable, the quantile $x(p)$ satisfies:

$$F_X(x(p)) = P(X < x(p)) = p \qquad (1.4.4.13)$$

A value $x(p)$ is also known as a *quantile of order p*. The quantile of order $p = 0.5$ is called the *median* of a distribution and, besides a mean, is the second measure of a "center" of distribution.

The remaining moments of a random variable, except $E(X)$, $D^2(X)$, are used rarely. For example, a central moment μ_3 is a basis for a construction of a *coefficient of skewness* of distribution of random variable X:

$$As = \frac{\mu_3}{D^3(X)} \qquad (1.4.4.14)$$

When $As = 0$, a distribution is *symmetric*, $As < 0$ ($As > 0$) characterize a *left-sided (right-sided) asymmetry*.

If a random variable has a small variance or standard deviation, we would expect most of the values to be grouped around the mean. Therefore, a probability' that a random variable assumes a value within a certain interval about a mean is greater than for a similar random variable with a larger standard deviation. The Russian mathematician P. L. Chebyshev gave a conservative estimate of a probability that a random variable assumes a value within k standard deviations of its mean for any real number k. This is formulated in the following theorem.

Chebyshev's Theorem
The probability that any random variable X will assume a value within k standard deviations of its mean is at least $(1 - 1/k^2)$. That is,

$$P[E(X) - k \cdot D(X) < X < E(X) + k \cdot D(X)] \geq 1 - 1/k^2 \qquad (1.4.4.15)$$

From Chebyshev's inequality it results the *rule of 3-standard deviations* which is very important and useful in practice of statistical analyses. A probability that a random variable X differ from its expected value by more than 3 standard deviations is very small:

$$P(|X - E(X)| \geq 3 \cdot D(X)) \leq \frac{1}{3^2} \qquad (1.4.4.16)$$

which means that such an observation practically does not happen. For this reason, if observed, are considered as outliers and are usually rejected in statistical analyses.
A standardization of a random variable
   Suppose X is a random variable with mean $E(X)$ and standard deviation $D(X) > 0$. Then a random variable U

$$U = \frac{X - E(X)}{D(X)} \qquad (1.4.4.17)$$

is a standardized form of random variable X. It can be shown that a standardized random variable U has a distribution with an expected value $E(U) = 0$ and variance $D^2(U) = 1$.

Example 1.4.4.1 The expected value and the variance of discrete random variable X from Example 1.4.2.1 are:

$$E(X) = -100 \cdot \frac{2}{3} + 100 \cdot \frac{1}{3} = -\frac{100}{3} \qquad (1.4.4.18)$$

$$D^2(X) = \left(-100 + \frac{100}{3}\right)^2 \cdot \frac{2}{3} + \left(100 + \frac{100}{3}\right)^2 \cdot \frac{1}{3} = \frac{800}{9} \qquad (1.4.4.19)$$

This result means that a player who plays over and over again will, on the average, lose approximately 33.33 \$. The spread of these loses around 33.33 \$ will be about $D(X) = 9.43$ \$.

The analogous values or the continuous random variable X from Example 1.4.3.1 are:

$$E(X) = \int\limits_{-\infty}^{+\infty} xf(x)dx = \int\limits_{0}^{1} xdx = \frac{1}{2} \qquad (1.4.4.20)$$

$$D^2(X) = \int\limits_{-\infty}^{+\infty} (x - E(X))^2 f(x)dx = \int\limits_{0}^{1} (x - \frac{1}{2})^2 dx = \frac{1}{12} \qquad (1.4.4.21)$$

1.4.5   Examples of Discrete Probability Distributions

Often, the observations generated by different statistical experiments have the same general type of behavior. Consequently, the discrete or continuous random variables associated with these experiments can be described by essentially the same probability distribution and, therefore, can be represented by a single formula. In fact, one needs only a handful of important probability distributions to describe many of the discrete or continuous random variables encountered in practice. Such a handful of distributions actually describe several real-life random phenomena. For instance, in an industrial example, when a sample of items selected from a batch of production is tested, the number of defective items in the sample usually can be modeled as a hypergeometric random variable.

The current and the next section present the commonly used discrete and continuous distributions with various examples.

Discrete Uniform Distribution
The simplest of all discrete probability distributions is the one where a random variable assumes each of its values with equal probability. Such a probability distribution is called a discrete uniform distribution. If the random variable X assumes the values $x_1, x_2, \ldots, x_k$ with equal probabilities, then the probability function of a discrete uniform distribution is given by:

$$P(X = x; k) = \frac{1}{k} \quad x = x_1, x_2, \ldots, x_k \qquad (1.4.5.1)$$

For example, when a fair die is rolled, each element of the sample space occurs with probability $P(X = x; 6) = 1/6$ for $x = 1, 2, 3, 4, 5, 6$.

Two-Point Distribution
The random variable X has a two-point distribution if its probability function is defined by the formula:

$$P(X = x_1) = p$$
$$P(X = x_2) = q$$

$$0 < p, q < 1, \quad p + q = 1 \tag{1.4.5.2}$$

where p, q are the parameters. The example of such distribution has been given in Example 1.4.2.1. The expectation and variance of this random variable are:

$$E(X) = px_1 + (1 - p)x_2 \quad D^2(X) = p(1 - p)(x_1 - x_2)^2$$

Binomial (Bernoulli) Distribution

An experiment often consists of repeated independent trials, each with the two possible outcomes that may be labelled as *success* or *failure*. The most obvious application deals with testing of items as they come off the assembly line, where each test or trial may indicate a defective or a nondefective item. The process is referred to as a *Bernoulli process*. Each trial is called a *Bernoulli trial*.

The number X of successes in n Bernoulli trials is called a *binomial random variable*. The probability distribution of this discrete random variable is called the binomial or Bernoulli distribution and its probability function is described by the formula:

$$P(X = x) = b(x; n, p) = \binom{n}{x} p^x q^{n-x} \tag{1.4.5.3}$$

where p is a probability of success. Its values are denoted by $b(x; n, p)$ (Fig. 1.4).
The expectation and variance of this random variable are:

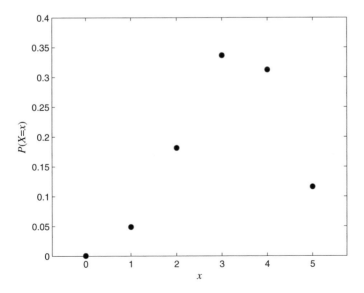

Fig. 1.4 Plot of probability function of Bernoulli random variable, $n = 5$, $p = 0.65$

$$E(X) = np \quad D^2(X) = np(1 - p) \tag{1.4.5.4}$$

Poisson Distribution

Experiments yielding the number of outcomes occurring during a given time interval are called *Poisson experiments*. Hence, a Poisson experiment can generate observations for a random variable X, representing for example a number of telephone calls per hour received by someone or a number of failures in a production process during one day, etc. The number of outcomes occurring in one-time interval in Poisson experiment is independent of the number that occurs in any other disjoint time interval (*no memory property*). The probability that a single outcome will occur during a very short time interval is proportional to its length and the probability that more than one outcome will occur in such a short time interval is negligible.

The probability distribution of the Poisson random variable X, representing the number x of outcomes occurring in a given time interval denoted by t, is given as:

$$P(X = x) = p(x; \lambda t) = e^{-\lambda t}\frac{(\lambda t)^x}{x!} \quad x = 0, 1, \dots \tag{1.4.5.5}$$

where λ $(\lambda > 0)$ is the average number of outcomes per unit time, and e $= 2.71828$.

The expectation and variance of Poisson random variables are:

$$E(X) = \lambda t \quad D^2(X) = \lambda t \tag{1.4.5.6}$$

If n is large and p is close to 0, the Poisson distribution can be used to approximate binomial probabilities: $b(x; n, p) = p(x; m)$ (with $m = np$). The Poisson distribution, like the binomial one, is often used in quality control or assurance. Sample plot of Poisson distribution is shown in Fig. 1.5.

Geometric Distribution

Let us consider an experiment where the properties are the same as those listed for a binomial experiment (a success with probability p and a failure with probability $q = 1 - p$). The only difference now is that independent trials will be repeated until a first success occur. Let's define the random variable X, being the number of trial on which the first success occurs, then, its probability function is described by the formula:

$$P(X = k) = pq^{k-1} \quad k = 1, 2, \dots \tag{1.4.5.7}$$

Since the successive terms constitute a geometric progression, it is customary to refer to this distribution as the *geometric distribution*. Sample plot of geometric distribution is shown in Fig. 1.6.

The expectation and variance of geometric random variable are:

$$E(X) = \frac{1}{p} \quad D^2(X) = \frac{1 - p}{p^2} \tag{1.4.5.8}$$

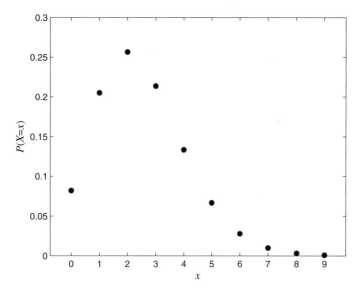

Fig. 1.5   Plot of probability function of Poisson random variable, $\lambda t = 2.5$

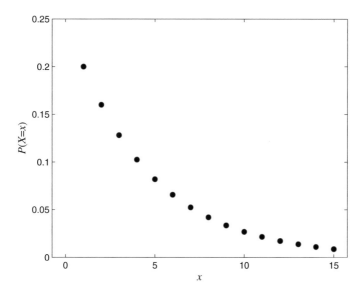

Fig. 1.6   Plot of probability function of geometric random variable, $p = 0.2$

Hypergeometric Distribution

A *hypergeometric experiment* can be described as follows: (1) a random sample of size n is selected without replacement from population of N items; (2) k of the N items may be classified as successes and $N - k$ are classified as failures. In general, we are interested in the probability of selecting x successes. The number X of successes of a hypergeometric experiment is called a *hypergeometric random variable*. Its probability function is given by the following formula:

$$P(X = x) = h(x; N, n, k) = \frac{\binom{k}{x} \cdot \binom{N-k}{n-x}}{\binom{N}{n}} \qquad (1.4.5.9)$$

$$x = 0, 1, \dots, \min(n, k)$$

The expectation and variance of hypergeometric random variable are:

$$E(X) = \frac{n \cdot k}{N} \qquad (1.4.5.10)$$

$$D^2(X) = \frac{N-n}{N-1} \cdot n \cdot \frac{k}{N} \left(1 - \frac{k}{N}\right) \qquad (1.4.5.11)$$

The hypergeometric distribution is often applied in *acceptance sampling* Sample plot of hypergeometric distribution is shown in 1.7.

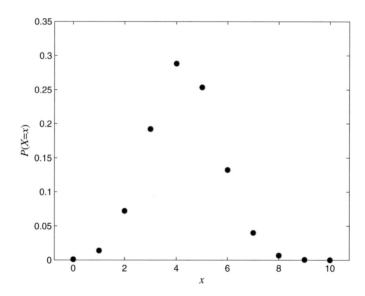

Fig. 1.7 Plot of probability function of a hypergeometric random variable N = 35, k = 15, n = 10

1.4.6 Examples of Continuous Probability Distributions

Normal Distribution

The most important, continuous probability distribution in the entire field of statistics is the normal distribution. Its plot, called the *normal curve*, is the bell-shaped curve (see Fig. 1.8), which describes approximately many phenomena that occur in nature, industry, or research. The normal distribution is often referred to as the *Gaussian distribution*, in honour of Karl Friedrich Gauss (1777–1855), who also derived its equation from a study of errors in repeated measurements of the same quantity. The mathematical equation for the probability distribution of the normal random variable depends upon the two parameters m and σ, its mean and standard deviation. Hence, we denote the values of the density as $N(x; m, \sigma)$. The probability density function has the form:

$$f(x) = N(x; m, \sigma) = \frac{1}{\sigma\sqrt{2\pi}} \exp\left(-\frac{(x-m)^2}{2\sigma^2}\right) \qquad (1.4.6.1)$$

It can be shown that the two parameters are related to the expected value and the variance:

$$E(X) = m \quad D^2(X) = \sigma^2 \qquad (1.4.6.2)$$

The normal distribution density plot has the following properties:

- It is symmetrical with respect to the line $x = m$

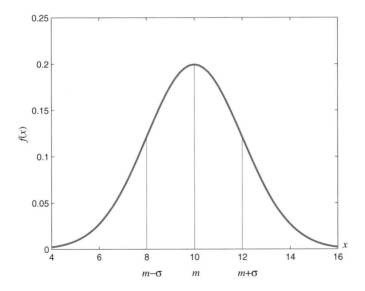

Fig. 1.8 Plot of normal probability density N (x; 10, 2)

- It reaches maximum equal to $\frac{1}{\sigma\sqrt{2\pi}}$ for $x = m$
- Its arms have inflection points for $x = m - \sigma$ and $x = m + \sigma$ (left and right concavity).

Under certain conditions the normal distribution provides a good continuous approximation to a binomial and hypergeometric distributions. Moreover, the limiting distribution of sample average is normal which provides a broad base for statistical inference.

For the normal curve, the probability that the random variable X assumes a value between the values x_1, x_2 is:

$$P(x_1 < X < x_2) = \frac{1}{\sigma\sqrt{2\pi}} \int_{x_1}^{x_2} \exp\left(-\frac{(x-m)^2}{2\sigma^2}\right) dx \qquad (1.4.6.3)$$

The difficulty encountered in solving integrals of normal density functions necessitates the tabulation of normal curve areas for quick reference. It would be a hopeless task to attempt to set up separate tables for every conceivable values of m, σ. Fortunately, by means of the standardisation

$$Z = \frac{X - m}{\sigma} \qquad (1.4.6.4)$$

we obtain the *standard normal distribution* (denoted as $N(x; 0, 1)$) with mean 0 and variance 1. The probability density and cumulative distribution functions of $N(x; 0, 1)$ are:

$$\varphi(x) = \frac{1}{\sqrt{2\pi}} \exp\left(-\frac{x^2}{2}\right) \quad \Phi(x) = \int_{-\infty}^{x} \varphi(t)dt \qquad (1.4.6.5)$$

We have now reduced the required number of tables of normal-curve areas to one, that of the standard normal distribution which is tabulated.

We will now provide three important distributions related to the normal distribution:

- chi-square χ^2
- Student's t
- F-Snedecor.

Chi-square Distribution (χ^2)
Let's $U_1, \ldots, U_k$ be independent random variables, each of which has a standard normal distribution $N(x; 0, 1)$. The random variable Y:

$$Y = \sum_{i=1}^{k} U_i^2 \tag{1.4.6.6}$$

has a *chi-square distribution* with k degrees of freedom $(chi(k))$. The density function of this distribution is expressed by the formula:

$$f(x) = \frac{x^{\frac{k}{2}-1}e^{-\frac{x}{2}}}{2^{\frac{k}{2}}\Gamma\left(\frac{k}{2}\right)} \tag{1.4.6.7}$$

where $\Gamma(.)$ means the *gamma function* defined by the formula (1.4.6.8). Sample plot of chi-square distribution is shown in Fig. 1.9.

$$\Gamma(p) = \int_0^{+\infty} x^{p-1} \cdot e^{-x}dx \quad \text{for } p > 0 \tag{1.4.6.8}$$

Student's t-Distribution
Let U and Y be the independent random variables with normal $N(x; 0, 1)$ and $chi(k)$ distributions, respectively. The random variable:

$$T = \frac{U}{\sqrt{\frac{Y}{k}}} \tag{1.4.6.9}$$

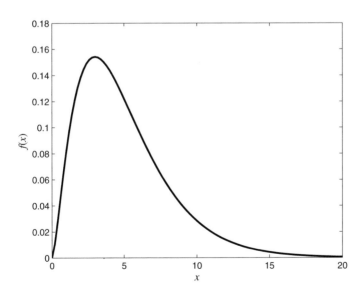

Fig. 1.9 Plot of chi-square probability density, k $= 5$

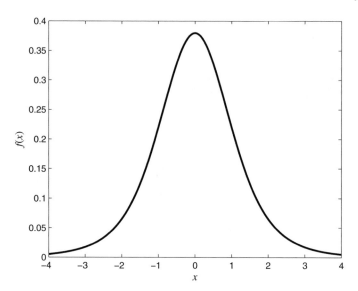

Fig. 1.10 Plot of Student's t probability density, $k = 5$

has a *Student's t-distribution* with k degrees of freedom. The density function of the Student's t-distribution is expressed by the formula:

$$f(x) = \frac{\Gamma\left(\frac{k+1}{2}\right)}{\sqrt{k\pi} \cdot \Gamma\left(\frac{k}{2}\right)} \cdot \left(1 + \frac{x^2}{k}\right)^{-\frac{k+1}{2}} \qquad (1.4.6.10)$$

where $\Gamma(.)$ means the gamma function defined above. The sample plot of Student's t-distribution is shown in Fig. 1.10.

F-Snedecor Distribution

Let U and V be two independent random variables with $chi(k_{nom})$ and $chi(k_{den})$ distributions, respectively. The random variable:

$$F = \frac{U/k_{nom}}{V/k_{den}} \qquad (1.4.6.11)$$

has *F-Snedecor* distribution with k_{nom} and k_{den} degrees of freedom of nominator and denominator, respectively (F in honour of English statistician R. Fisher). The sample plot of F-Snedecor's distribution is shown in Fig. 1.11 (the complicated formula is not included).

Exponential Distribution

The probability density function of a continuous random variable X with exponential distribution with parameter a is expressed by the formula:

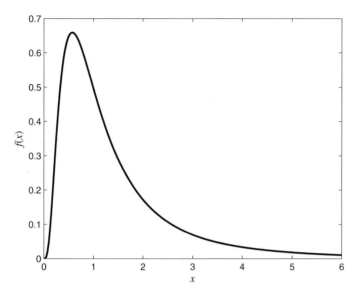

Fig. 1.11 Plot of F-Snedecor's probability density, number of degrees of freedom of the nominator—10, denominator—5

$$f(x; a) = \begin{cases} 0 & x < 0 \\ ae^{-ax} & x \geq 0 \end{cases} \qquad (1.4.6.12)$$

The expectation and variance of this distribution are defined by the formula (1.4.6.13). The sample plot of exponential distribution is shown in Fig. 1.12.

$$E(X) = \frac{1}{a} \quad D^2(X) = \frac{1}{a^2} \qquad (1.4.6.13)$$

Exponential distribution is a special case (p = 1) of *gamma distribution* (with scale a and shape p parameters) characterized by the following probability density function:

$$f(x; a, p) = \frac{a^p}{\Gamma(p)} x^{p-1} e^{-ax} \qquad (1.4.6.14)$$

Continuous Uniform Distribution

One of the simplest continuous distributions in all of statistics is the continuous uniform distribution. This distribution is characterized by a density function that is "flat" and, thus, the probability is uniform in a closed interval, say [a, b]. The probability density function of the continuous uniform random variable X on the interval [a, b] is:

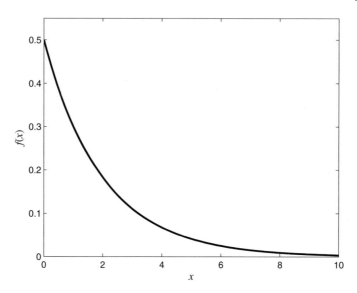

Fig. 1.12 Plot of exponential probability density, a = 1/2

$$f(x) = \begin{cases} \frac{1}{b-a} & x \in [a, b] \\ 0 & x \notin [a, b] \end{cases} \tag{1.4.6.15}$$

The cumulative distribution function of that random variable is given by the formula:

$$F(x) = \begin{cases} 0 & x < a \\ \frac{x-a}{b-a} & x \in [a, b] \\ 1 & x > b \end{cases} \tag{1.4.6.16}$$

The notation $X \sim unif(a, b)$ states that a random variable X has a uniform distribution on the interval [a, b]. It can be shown that the expectation and variance of such random variable X are:

$$E(X) = \frac{a+b}{2} \quad D^2(X) = \frac{(b-a)^2}{12} \tag{1.4.6.17}$$

The plot of probability density and cumulative distribution functions of the uniform random variable on the interval [0, 1] have been shown in the Example 1.4.3.1.

1.5   Two or More Random Variables

In real life, we are often interested in several random variables that are related to each other. For example, suppose that we choose a random family, and we would like to study the number of people in the family (X_1) and the household income (X_2). Both X_1 and X_2 are random variables, and we suspect that they are dependent. Formally, a vector:

$$(X_1, X_2, \ldots, X_n) \tag{1.5.1}$$

is called an *n-dimensional random variable (n-dimensional random vector)* where each X_i, for $i = 1, \ldots, n$ is defined on the same probability space (Ω, Z, P).

Given the random variables X_1, X_2 that are defined in a common probability space, the *joint probability distribution* for X_1, X_2 is a probability distribution that gives the probability that each of X_1, X_2 falls in any particular range or discrete set of values specified for that variable. In other words, the probability distribution that defines simultaneous behavior of two or more random variables is called a *joint probability distribution*.

In this chapter, we develop tools to study the joint distributions of random variables and we focus only on two random variables, denoted here as (X, Y) (*bivariate case*). The extension to *multivariate case* ($n > 2$ random variables) is, then, straightforward.

Depending on whether both random variables X and Y are discrete or continuous, we talk about a *discrete* or *continuous random vector*. A *mixed random vector* is also possible.

The joint probability distribution can be expressed either in terms of a *joint cumulative distribution function* or in terms of a *joint probability density function* (in the case of continuous variables) or *joint probability mass function* (in the case of discrete variables). These, in turn, can be used to find two other types of distributions: the *marginal distribution* giving the probabilities for any one of the variables with no reference to any specific ranges of values for the other variables, and the *conditional probability distribution* giving the probabilities for any subset of the variables conditional on particular values of the remaining variables.

A *joint cumulative distribution function* of two random variables X and Y is defined as:

$$F_{XY}(x, y) = P(X < x, Y < y) \tag{1.5.2}$$

The value of a cumulative distribution at (x, y) is the probability that a random variable X assumes a value less than x and a random variable Y assumes a value less than y (comma means "and").

1.5.1 Two Discrete Random Variables

Let the two random variables X and Y be defined on the common probability space of an experiment. A discrete random vector (X, Y) assumes a finite or countable number of pair of values $(x_i, y_j) \in R^2$. The joint probability distribution of a random vector (X, Y) is determined by the *joint probability mass function* (pmf) $P(x, y)$ defined for each pair of values (x_i, y_j) of a vector (X, Y) by:

$$P(\{X = x_i\} \cap \{Y = y_j\})$$
$$= P(\{X = x_i, Y = y_j\}) = p(x_i, y_j) = p_{ij} \qquad (1.5.1.1a)$$

It must be the case that:

$$p_{ij} \geq 0 \quad \text{and} \quad \sum_j \sum_i p_{ij} = 1 \qquad (1.5.1.1b)$$

The joint pmf of discrete random vector (X, Y) describes how much probability mass is placed on each possible pair of values (x, y). Now let A be any set consisting of pairs of values (x, y). Then, the probability $P(X, Y) \in A$ is obtained by summing the joint pmf over pairs in A:

$$P(X, Y) \in A = \sum_{(x_i, y_j) \in A} \sum p(x_i, y_j) \qquad (1.5.1.2)$$

The cumulative distribution function of a discrete vector (X, Y) is:

$$F(x, y) = P(X < x, Y < y) = \sum_{x_i < x} \sum_{y_i < y} p_{ij} \qquad (1.5.1.3)$$

The joint pmf of discrete random vector (X, Y) can be presented in the accompanying *joint probability table* (see Table 1.4) in which the first row and column

Table 1.4 The joint probability table of (X, Y) with marginal distributions

X	Y				
	y_1	y_2	...	y_k	$p_{i.} = \sum_{j=1}^{k} p_{ij}$
x_1	p_{11}	p_{12}	...	p_{1k}	$p_{1.}$
x_2	p_{21}	p_{22}	...	p_{2k}	$p_{2.}$
...	...	...	...	...	...
x_r	p_{r1}	p_{r2}	...	p_{rk}	$p_{r.}$
$p_{.j} = \sum_{i=1}^{r} p_{ij}$	$p_{.1}$	$p_{.2}$	...	$p_{.k}$	$\sum_{i=1}^{r} \sum_{j=1}^{k} p_{ij} = 1$

contain the values of variables Y and X, respectively. In the central part of the table there are the probabilities p_{ij} of the joint probability mass function.

Marginal Distribution

Once the joint pmf of the two variables X and Y is available, it is straight-forward to obtain the distribution of just one of these variables. For any possible value x of X, the probability $P(\{X = x\})$ results from holding x fixed and summing the joint pmf $P(x, y)$ over all y for which the pair (x, y) has a positive probability mass. The same strategy applies to obtaining the distribution $P(\{Y = y\})$ of Y.

The *marginal probability mass function* of X is given by:

$$P(x_i) = P(\{X = x_i\}) = p_{i.} = \sum_{j=1}^{k} p_{ij} \quad i = 1, \ldots, r \qquad (1.5.1.4)$$

where

$$\sum_{i} p_{i.} = 1 \qquad (1.5.1.5)$$

Similarly, the *marginal probability mass function* of Y is:

$$P(\{Y = y_j\}) = p_{.j} = \sum_{i=1}^{r} p_{ij} \quad j = 1, \ldots, k \qquad (1.5.1.6)$$

where

$$\sum_{j} p_{.j} = 1 \qquad (1.5.1.7)$$

The use of the word marginal here is a consequence of the fact that if the joint pmf is displayed in a rectangular table as in Table 1.4, then, the row totals give the marginal pmf of X and the column totals give the marginal pmf of Y.

The marginal distribution of a random vector (X, Y) can also be determined by *marginal cumulative distribution functions:*

$$F_X(x) = P(X < x, Y < \infty) \qquad (1.5.1.8)$$

$$F_Y(y) = P(X < \infty, Y < y) \qquad (1.5.1.9)$$

If (X, Y) is of discrete type, then, the above expressions are reduced to the following:

$$F_X(x) = P(X < x, Y < \infty) = \sum_{x_i < x} p_{i.} \qquad (1.5.1.10)$$

$$F_Y(y) = P(X < \infty, Y < y) = \sum_{y_j < y} p_{\cdot j} \qquad (1.5.1.11)$$

Conditional Distribution

Given the two jointly distributed discrete random variables X and Y, the conditional probability distribution of Y given X, is the probability distribution of Y when X is known to be a particular value. Formally, *the conditional probability mass function of variable X given $Y = y$* is defined as follows (see Table 1.4):

$$P(\{X = x_i | Y = y_j\})$$
$$= \frac{P(\{X = x_i \cap Y = y_j\})}{P(\{Y = y_j\})} = \frac{p_{ij}}{p_{\cdot j}} \quad i = 1, \ldots r \qquad (1.5.1.12)$$

Similarly, the *conditional probability mass function of variable Y given $X = x$* is defined as:

$$P(\{Y = y_j | X = x_i\})$$
$$= \frac{P(\{Y = y_j \cap X = x_i\})}{P(\{X = x_i\})} = \frac{p_{ij}}{p_{i\cdot}} \, j = 1, \ldots, k \qquad (1.5.1.13)$$

Conditional distributions of random vector (X, Y) can also be determined by *conditional cumulative distribution functions*:

$$F(x|y) = P(X < x | Y = y) \qquad (1.5.1.14)$$

$$F(y|x) = P(Y < y | X = x) \qquad (1.5.1.15)$$

If (X, Y) is of a discrete type, the above expressions are reduced to:

$$F(x|y) = P(X < x | Y = y) = \sum_{x_i < x} P(X = x_i | Y = y) \qquad (1.5.1.16)$$

$$F(y|x) = P(Y < y | X = x) = \sum_{y_j < y} P(Y = y_j | X = x) \qquad (1.5.1.17)$$

Below we present the example illustrating the calculation of the joint, marginal and conditional distributions of random vector (X, Y).

Example 1.5.1.1 Let us consider the experiment of simultaneously tossing a coin and rolling a die. The result of the coin toss is a random variable X assuming the values 0 and 1 with equal probabilities of 1/2. The result of the die roll is a random variable Y that assumes values 1, 2, 3, 4, 5, 6 with equal probabilities of 1/6. The

Table 1.5 The joint and marginal distributions of vector (X, Y) from Example 1.5.1.1

X	Y						
	1	2	3	4	5	6	$p_i.$
0	$\frac{1}{12}$	$\frac{1}{12}$	$\frac{1}{12}$	$\frac{1}{12}$	$\frac{1}{12}$	$\frac{1}{12}$	$\frac{6}{12}$
1	$\frac{1}{12}$	$\frac{1}{12}$	$\frac{1}{12}$	$\frac{1}{12}$	$\frac{1}{12}$	$\frac{1}{12}$	$\frac{6}{12}$
$p.j$	$\frac{2}{12}$	$\frac{2}{12}$	$\frac{2}{12}$	$\frac{2}{12}$	$\frac{2}{12}$	$\frac{2}{12}$	1

result of such simultaneous tossing and rolling can be described by a random vector (X, Y) that can assume the following pairs of values:

(0, 1) (0, 2) (0, 3) (0, 4) (0, 5) (0, 6) (1, 1) (1, 2) (1, 3) (1, 4) (1, 5) (1, 6)

The probability of occurring of each such pair (event) is:

$$P\left(X = x_i, Y = y_j\right) = \frac{1}{2} \cdot \frac{1}{6} = \frac{1}{12} \quad i = 1, 2 \quad j = 1, \ldots, 6 \tag{1.5.1.18}$$

The joint distribution of (X, Y) has yet been defined and is presented in the Table 1.5. The last column and row of this table show the values of the probability mass function of the two marginal distributions of (X, Y).

Let us now determine the conditional distributions of the vector (X, Y). There are 6 conditional distributions of the random variable X given the variable Y, each containing the two values of the random variable X at a particular value of variable $Y = 1, \ldots, 6$:

$$P(X = 0|Y = 1) = \frac{1/12}{2/12} = \frac{1}{2} \quad P(X = 1|Y = 1) = \frac{1/12}{2/12} = \frac{1}{2}$$

$$P(X = 0|Y = 2) = \frac{1/12}{2/12} = \frac{1}{2} \quad P(X = 1|Y = 2) = \frac{1/12}{2/12} = \frac{1}{2}$$

$$P(X = 0|Y = 3) = \frac{1/12}{2/12} = \frac{1}{2} \quad P(X = 1|Y = 3) = \frac{1/12}{2/12} = \frac{1}{2}$$

$$P(X = 0|Y = 4) = \frac{1/12}{2/12} = \frac{1}{2} \quad P(X = 1|Y = 4) = \frac{1/12}{2/12} = \frac{1}{2}$$

$$P(X = 0|Y = 5) = \frac{1/12}{2/12} = \frac{1}{2} \quad P(X = 1|Y = 5) = \frac{1/12}{2/12} = \frac{1}{2}$$

$$P(X = 0|Y = 6) = \frac{1/12}{2/12} = \frac{1}{2} \quad P(X = 1|Y = 6) = \frac{1/12}{2/12} = \frac{1}{2}$$

There are two conditional distributions of the random variable Y given the variable X. The first distribution, containing 6 values of the variable Y at the value $X = 0$:

$$P(Y = 1|X = 0) = \frac{1/12}{6/12} = \frac{1}{6}$$

$$\cdots$$

$$P(Y = 6|X = 0) = \frac{1/12}{6/12} = \frac{1}{6}$$

The second distribution, also with 6 values of Y at a value, $X = 1$ and so on:

$$P(Y = 1|X = 1) = \frac{1/12}{6/12} = \frac{1}{6}$$

$$\cdots$$

$$P(Y = 6|X = 1) = \frac{1/12}{6/12} = \frac{1}{6}$$

1.5.2 Two Continuous Random Variables

The continuous random vector (X, Y) assumes the infinite and uncountable number of pairs of values $(x_i, y_j) \in R^2$. The probability that the pair (X, Y) of continuous random variables falls in a two-dimensional set A (such as a rectangle) is obtained by integrating a function called the joint density function.

Formally, let X and Y be continuous random variables. A *joint probability density function* $f(x, y)$ for these two variables is a function satisfying conditions:

$$f(x, y) \geq 0 \tag{1.5.2.1}$$

$$\int_{-\infty}^{+\infty} \int_{-\infty}^{+\infty} f(x, y)dxdy = 1 \tag{1.5.2.2}$$

A *joint cumulative distribution* function of a continuous random vector (X, Y) is given by the formula:

$$F(x, y) = P(X < x, Y < y) = \int_{-\infty}^{x} \int_{-\infty}^{y} f(x, y)dxdy \tag{1.5.2.3}$$

Marginal Distribution

The marginal pdf of each variable can be obtained in a manner analogous to what we did in the case of two discrete variables. The marginal pdf of X at the value x results from holding x fixed in the pair (x, y) and integrating the joint pdf over y. Integrating the joint pdf with respect to x gives the marginal pdf of Y. The *marginal probability density functions* of X and Y, denoted by $f_X(x)$ and $f_Y(y)$, respectively, are given by:

$$f_X(x) = \int\limits_{-\infty}^{+\infty} f(x, y)dy \tag{1.5.2.4}$$

$$f_Y(y) = \int\limits_{-\infty}^{+\infty} f(x, y)dx \tag{1.5.2.5}$$

The formulas for the *marginal cumulative distribution functions* in the case of a continuous vector (X, Y) take the form:

$$F_X(x) = \int\limits_{-\infty}^{x} f_X(x)dx \quad F_Y(y) = \int\limits_{-\infty}^{y} f_Y(y)dy \tag{1.5.2.6}$$

Conditional Distribution

The conditional probability density function of the variable X given $Y = y$ is defined as follows:

$$f(x|y) = \frac{f(x, y)}{f_Y(y)} \tag{1.5.2.7}$$

Similarly, the *conditional probability density function* of the variable Y given $X = x$ is defined as:

$$f(y|x) = \frac{f(x, y)}{f_X(x)} \tag{1.5.2.8}$$

Conditional distributions of random continuous vector (X, Y) can also be determined by the *conditional cumulative distribution functions*:

$$F(x|y) = \int\limits_{-\infty}^{x} f(x|y)dx \quad F(y|x) = \int\limits_{-\infty}^{y} f(y|x)dy \tag{1.5.2.9}$$

Example 1.5.2.1 Let the joint probability density function of the random vector (X, Y) is specified by the formula:

$$f(x, y) = \begin{cases} (x - y)^2 & -1 \le x \le 1, -1 \le y \le 1 \\ 0 & otherwise \end{cases} \tag{1.5.2.10}$$

Let $-1 \le x \le 1$. We have

$$f_X(x) = \int\limits_{-\infty}^{+\infty} f(x, y)dy = \int\limits_{-1}^{1} (x - y)^2 dy = \int\limits_{-1}^{1} (x^2 - 2xy + y^2)dy$$

$$= \left[x^2 y - xy^2 + \frac{y^3}{3} \right]\Big|_{-1}^{+1} = 2(x^2 + \frac{1}{3}) \qquad (1.5.2.11)$$

Thus, the marginal probability density function of X has the form:

$$f_X(x) = \begin{cases} 2(x^2 + \frac{1}{3}) & -1 \le x \le 1 \\ 0 & otherwise \end{cases} \qquad (1.5.2.12)$$

Due to the symmetry, the marginal probability density function of the variable Y has the same form:

$$f_Y(y) = \begin{cases} 2(y^2 + \frac{1}{3}) & -1 \le y \le 1 \\ 0 & otherwise \end{cases} \qquad (1.5.2.13)$$

Now, let $-1 \le y \le 1$. The conditional probability density function of random variable X, given $Y = y$:

$$f(x|y) = \frac{f(x, y)}{f_Y(y)} = \frac{(x - y)^2}{2(y^2 + \frac{1}{3})} \quad if \ \ x \in [-1, 1] \qquad (1.5.2.14)$$

$$f(x|y) = 0 \ \ if \ \ x \notin [-1, 1] \qquad (1.5.2.15)$$

Similarly, we can determine the conditional distribution of the random variable Y given $X = x$.

1.5.3   Independent Random Variables

In many situations, information about the observed value of one of the two random variables X and Y gives information about the value of the other variable, thus, there is a dependence between the two variables.

Two random variables X and Y are said to be *independent* if for each pair of their values (x_i, y_j) the following holds:

$$P(\{X = x_i, Y = y_j\}) = P(X = x_i) \cdot P(Y = y_j) \qquad (1.5.3.1)$$

or equivalently

$$p_{ij} = p_{i.} \cdot p_{.j} \qquad (1.5.3.2)$$

when X and Y are discrete and:

$$f(x, y) = f_X(x) \cdot f_Y(y) \qquad (1.5.3.3)$$

when X and Y are continuous.

If the above conditions are not satisfied, then X and Y are said to be *dependent*.

Example 1.5.3.1 The random variables X and Y from Example 1.5.2.1 are dependent:

$$f(x, y) = (x - y)^2 \neq f_X(x) \cdot f_Y(y) = 4\left(x^2 + \frac{1}{3}\right) \cdot \left(y^2 + \frac{1}{3}\right) \qquad (1.5.3.4)$$

1.5.4 Expectations, Covariance and Correlation

Let X and Y be jointly distributed random variables with pmf p_{ij} or pdf $f(x, y)$ according to whether the variables are discrete or continuous. Then, their expected values (expectations) are given by:

$$E(X) = \sum_{i=1}^{r} x_i \cdot p_{i.} \quad E(Y) = \sum_{j=1}^{k} y_j \cdot p_{.j} \qquad (1.5.4.1)$$

for discrete vector (X, Y) and:

$$E(X) = \int_{-\infty}^{+\infty} \int_{-\infty}^{+\infty} x f(x, y) dx dy \quad E(Y) = \int_{-\infty}^{+\infty} \int_{-\infty}^{+\infty} y f(x, y) dx dy \qquad (1.5.4.2)$$

for a continuous random vector (X, Y). These are the expectations in the marginal distribution of X and Y, respectively.

When two random variables X and Y are not independent, it is frequently of interest to assess how strongly they are related to one another.

The covariance between the random variables X and Y is defined by the following formula:

$$\mathrm{cov}(X, Y) = \sigma_{XY} = E[(X - E(X))(Y - E(Y))] =$$

$$= \begin{cases} \sum_i \sum_j (x_i - E(X))(y_i - E(Y)) \cdot p_{ij} & \text{discrete} \\ \int_{-\infty}^{+\infty} \int_{-\infty}^{+\infty} (x - E(X))(y - E(Y)) f(x, y) dx dy & \text{continuous} \end{cases}$$

$$(1.5.4.3)$$

The following shortcut formula for $\mathrm{cov}(X, Y)$ simplifies the computations:

$$\mathrm{cov}(X, Y) = E(X \cdot Y) - E(X) \cdot E(Y) \tag{1.5.4.4}$$

For a strong positive relationship, $\mathrm{cov}(X, Y)$ should be quite positive while for a strong negative—quite negative. If X and Y are not strongly related, covariance will be near 0.

The defect of covariance is that its computed value depends critically on the units of measurement. Ideally, the choice of units should have no effect on a measure of strength of relationship. This is achieved by scaling the covariance.

Correlation coefficient between variables X and Y is used, which is a normalized covariance:

$$\rho = \frac{\mathrm{cov}(X, Y)}{\sqrt{D^2(X) \cdot D^2(Y)}} = \frac{\mathrm{cov}(X, Y)}{D(X) \cdot D(Y)} \tag{1.5.4.5}$$

where $D(X)$ and $D(Y)$ are the standard deviations in the marginal distributions of X and Y, respectively.

Correlation coefficient between the variables X and Y satisfies the condition:

$$-1 \leq \rho \leq +1 \tag{1.5.4.6}$$

The correlation coefficient ρ is actually a measure of the degree of linear relationship only, which results from the following theorem.

Theorem 1.5.4.1

1. *If X and Y are independent, then $\rho = 0$, but $\rho = 0$ does not imply independence.*
2. *Correlation coefficient $\rho^2 = 1$ if and only if:*

$$P(Y = aX + b) = 1 \tag{1.5.4.7}$$

for some numbers a, b and a $\neq$ 0.

Only when the two variables are perfectly related in a linear manner will ρ be as positive or negative as it can be. A ρ less than 1 indicates only that the relationship is not completely linear. When $\rho = 0$, X and Y are said to be *uncorrelated*. Two variables could be uncorrelated but highly dependent because there is a strong nonlinear relationship.

In the case of a multidimensional random variable $X = (X_1, X_2, \ldots, X_n)$, the two most important parameters characterizing its distribution are: the *vector of expected values:*

$$\mu = E(X) = [E(X_1), E(X_2), \ldots, E(X_n)] \tag{1.5.4.8}$$

and the *covariance matrix*, defined as:

$$C = E[(X - EX)(X - EX)^T] =$$

$$= \begin{bmatrix} D^2(X_1) & \text{cov}(X_1, X_2) & \ldots & \text{cov}(X_1, X_n) \\ \text{cov}(X_2, X_1) & D^2(X_2) & \ldots & \text{cov}(X_2, X_n) \\ & & \ldots & \\ \text{cov}(X_n, X_1) & \text{cov}(X_n, X_2) & \ldots & D^2(X_n) \end{bmatrix} \qquad (1.5.4.9)$$

The diagonal elements are the variances $D^2(X_i)$ of the corresponding variable X_i, the remaining ones are the covariances $\text{cov}(X_i, X_j)$ of the corresponding pair of variables (X_i, X_j). The covariance matrix is symmetrical. If the random variables $X_1, \ldots, X_n$ forming the random vector are pairwise independent, the covariance matrix is a diagonal matrix.

Example 1.5.4.1 Let's calculate the covariance of random variables X and Y from Example 1.5.1.1. The expected values of the two marginal distributions are:

$$E(X) = 0 \cdot \frac{6}{12} + 1 \cdot \frac{6}{12} = \frac{6}{12} \qquad (1.5.4.10)$$

$$E(Y) = \frac{2}{12}(1 + 2 + 3 + 4 + 5 + 6) = \frac{42}{12} \qquad (1.5.4.11)$$

$$\text{cov}(X, Y) = \frac{1}{12}\{(0 - \frac{6}{12})[(1 - \frac{42}{12}) + \ldots + (6 - \frac{42}{12})] +$$

$$+ (1 - \frac{42}{12})[(0 - \frac{6}{12}) + \ldots + (6 - \frac{42}{12})]\}$$

$$= \frac{1}{12}[(1 - \frac{42}{12}) + \ldots + (6 - \frac{42}{12})](-\frac{6}{12} + \frac{6}{12}) = 0 \qquad (1.5.4.12)$$

It follows that the random variables X and Y are uncorrelated.

1.5.5   Examples of Multidimensional Distributions

k-dimensional Normal Distribution

The joint probability density function of k-dimensional normal distribution is expressed by the formula:

$$f(x) = (2\pi)^{-\frac{1}{2}k} |C|^{-\frac{1}{2}} \exp\left[-\frac{1}{2}(x - \mu)^T C^{-1}(x - \mu)\right] \qquad (1.5.5.1)$$

where

$$x = (x_1, x_2, \ldots, x_k)$$

μ the vector of expected values,
$|C|$ determinant of the covariance matrix C.

Polynomial Distribution

Consider n independent repetitions of some statistical experiment, which can result in one of k mutually exclusive events:

$$A_1, \ldots, A_k \tag{1.5.5.2}$$

The probabilities of these events are given by:

$$P(A_i) = p_i \quad i = 1, \ldots, k, \quad \sum_{i=1}^{k} p_i = 1 \tag{1.5.5.3}$$

Let's define a k-dimensional random vector:

$$(X_1, \ldots, X_k) \tag{1.5.5.4}$$

whose component X_i is a random variable denoting the number of occurrences of an event A_i in n independent experiments. Therefore, the realizations of the vector $(x_1, \ldots, x_k)$ satisfy the condition:

$$\sum_{i=1}^{k} x_i = n \tag{1.5.5.5}$$

A random vector $(X_1, \ldots, X_k)$ defined in this way has a *polynomial distribution* with a joint probability mass function specified by a formula:

$$P(X_1 = x_1, \ldots, X_k = x_k) = \frac{n!}{x_1! \ldots x_k!} p_1^{x_1} \ldots p_k^{x_k}$$

$$x_i = 0, 1, \ldots, n \quad i = 1, \ldots, k \tag{1.5.5.6}$$

The above formula determines the joint probability of an event that in n experiments event A_1 will happen x_1 times and the event A_2 will happen x_2 times, and so on.

1.5.6 Conditional Expectation

The function r which assigns to each individual value x the expected value of the conditional distribution of the variable Y given $X = x$:

$$r(x) = E(Y|X = x) \qquad (1.5.6.1)$$

is called the *type 1 regression function* oh Y given X. The values $E(Y|X = x)$ are called *conditional expectations*.

In an analogous way, one can define the *type 1 regression function* oh X given $Y = y$:

$$r(y) = E(X|Y = y) \qquad (1.5.6.2)$$

The regression function can be linear, quadratic, exponential, and so on. The regression function of independent random variables X and Y is a constant function.

The regression function $r(x)$ of the random variable Y given X, in case when these variables are correlated (and therefore stochastically dependent), enables prediction of the value of the variable Y based on the known value of the variable X.

Example 1.5.6.1 Let us calculate the expected values of conditional distributions of the random variable X given the variable Y in the distribution of the vector from the Example 1.5.1.1.

$$E(X|Y = 1) = 0 \cdot \frac{1}{2} + 1 \cdot \frac{1}{2} = \frac{1}{2} \qquad (1.5.6.3)$$

$$E(X|Y = 2) = 0 \cdot \frac{1}{2} + 1 \cdot \frac{1}{2} = \frac{1}{2} \qquad (1.5.6.4)$$

$$E(X|Y = 3) = 0 \cdot \frac{1}{2} + 1 \cdot \frac{1}{2} = \frac{1}{2} \qquad (1.5.6.5)$$

$$E(X|Y = 4) = 0 \cdot \frac{1}{2} + 1 \cdot \frac{1}{2} = \frac{1}{2} \qquad (1.5.6.6)$$

$$E(X|Y = 5) = 0 \cdot \frac{1}{2} + 1 \cdot \frac{1}{2} = \frac{1}{2} \qquad (1.5.6.7)$$

$$E(X|Y = 6) = 0 \cdot \frac{1}{2} + 1 \cdot \frac{1}{2} = \frac{1}{2} \qquad (1.5.6.8)$$

As you can see, the obtained regression function $r(y)$ of the random variable X given Y is a constant function:

$$r(y) = 1/2 \qquad (1.5.6.9)$$

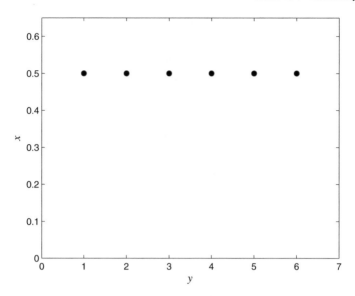

Fig. 1.13 Plot of regression of X versus Y for Example 1.5.6.1

for $y = 1, 2, 3, 4, 5, 6$ (Fig. 1.13).

This is due to the fact that the random variables X and Y are independent, which is easy to check, because for all the pairs (x_i, y_j) of values in the Table 1.5 the equality holds:

$$p_{ij} = p_{i.} \cdot p_{.j} \qquad (1.5.6.10)$$

which follows from:

$$\frac{1}{12} = \frac{6}{12} \cdot \frac{2}{12} \qquad (1.5.6.11)$$

1.6   Limit Theorems

In this section, we will discuss the important theorems in probability, often called the *limit theorems*, namely the laws of large numbers (LLN) and the central limit theorem (CLT). Since the limit theorems concern themselves with an asymptotic behavior of a sequence of random variables, we should first give the necessary definitions of different types of convergence of random variables.

We begin by recalling some defnitions pertaining to convergence of a sequence of real numbers. Let $\{x_n\}$ be a real-valued sequence. We say that the sequence $\{x_n\}$ converges to some $x \in R$ (Rudin 1976, p.47) if there exists an $n_0 \in N$ such that for all $\varepsilon > 0$,

$$|x_n - x| < \varepsilon, \quad \forall\, n \geq n_0 \tag{1.6.1}$$

We denote that fact as $\lim\limits_{n\to\infty} x_n = x$, or, shortly $x_n \to x$. Since random variables satisfy the measurability property, hence, the simplest notion of convergence of a sequence of random variables is defined in a fashion similar to that for regular functions. Let (Ω, Z, P) be a probability space and let:

$$X_1, X_2, \ldots, X_n \tag{1.6.2}$$

be a sequence $\{X_n\}$ of real-valued random variables defined on this probability space. We assume that these random variables are mutually independent and identically distributed which is usually abbreviated as *i.i.d.*

Point-wise convergence or sure convergence

A sequence of random variables $\{X_n\}$ is said to converge point-wise or surely to X if:

$$X_n(\omega) \to X(\omega) \quad \forall \omega \in \Omega \tag{1.6.3}$$

Note that for a fixed ω, $\{X_n(\omega)\}$ is a sequence of real numbers. Hence, the convergence for this sequence is same as the one in Definition (1.6.1). Since this notion is too strict for most practical purposes, and does not consider the probability measure, we define other notions.

Almost sure convergence or convergence with probability 1

A sequence of random variables $\{X_n\}$ is said to converge almost surely or with probability 1 to X if;

$$P(\{\omega : X_n(\omega) \to X(\omega)\}) = 1 \tag{1.6.4}$$

It is shortly denoted by $X_n \overset{w.p.1}{\to} X$. Almost sure convergence demands that the set of ω's where the random variables converge have a probability one. In other words, this definition gives the random variables "freedom" not to converge on a set of zero measure!

Convergence in probability

A sequence of random variables $\{X_n\}$ is said to converge in probability to X if:

$$\lim_{n\to\infty} P(|X_n - X| > \varepsilon) = 0 \quad \forall\, \varepsilon > 0 \tag{1.6.5}$$

It is shortly denoted by $X_n \overset{P}{\to} X$. As seen from the above definition, this notion concerns itself with the convergence of a sequence of probabilities!

Convergence in distribution or weak convergence

A sequence of random variables $\{X_n\}$ is said to converge in distribution to X if:

$$\lim_{n\to\infty} F_{X_n}(x) = F_X(x) \tag{1.6.6}$$

at all points of continuity of $F_X(x)$. It is shortly denoted by $X_n \overset{D}{\rightarrow} X$.

Let $X_1 X_2, \ldots$ be i.i.d random variables with $E(|X_i|) = m_i < \infty$. In general, we say that a sequence $\{X_n\}$ of random variables satisfies the **weak law of large numbers** (**WLLN**) if the convergence is according to probability:

$$\frac{1}{n} \sum_{i=1}^{n} (X_i - m_i) \overset{P}{\rightarrow} 0 \tag{1.6.7}$$

A sequence $\{X_n\}$ of random variables satisfies the **strong law of large numbers** (**SLLN**) if convergence is with probability 1:

$$\frac{1}{n} \sum_{i=1}^{n} (X_i - m_i) \overset{w.p.\ 1}{\rightarrow} 0 \tag{1.6.8}$$

The sense of convergence in the weak laws of large numbers is convergence in probability. The strong laws of large numbers, as the name suggests, assert the stronger notion of almost sure convergence.

Let's illustrate the above introduced LLN by the examples. Denote by p a probability of success in n Bernoulli trials, X_i are random variables with a zero-one distribution:

$$P(X_i = 1) = p \quad P(X_i = 0) = 1 - p \tag{1.6.9}$$

Then for every $\varepsilon > 0$:

$$\lim_{n \to \infty} P\left(\left| \frac{S_n}{n} - p \right| < \varepsilon \right) = 1 \tag{1.6.10}$$

where $S_n = \sum_{i=1}^{n} X_i$ is a number of successes in n trials.

Thus, after sufficiently many trials, the observed relative frequency S_n/n of success will, with a high degree of certainty, stay within a given ε of the probability p of success. As we can see, this convergence is according to a probability $\frac{S_n}{n} \overset{P}{\rightarrow} p$. This law of large numbers was discovered and proved by Jacob Bernoulli (1655-1705), therefore, often it is called **Bernoulli LLN**. It turns out that the same assumptions also coincide with an almost sure convergence, i.e. $\frac{S_n}{n} \overset{w.p.\ 1}{\rightarrow} p$. Let's consider a fair coin toss as Bernoulli trial. When a coin is flipped once, the theoretical probability that the outcome will be head (a success) is equal to $p = 1/2$. Therefore, according to the LLN, the proportion of heads in a "large" number of coin flips "should be" roughly 1/2. In particular, the proportion of heads after n flips will almost surely converge to 1/2 as n approaches infinity.

We will now present the **central limit theorem** (**CLT**), which establishes that in some situations, the distribution of properly normalized mean of independent random variables tends toward a standard normal distribution even if the original variables themselves are not normally distributed.

More formally, let $\{X_n\}$ be a sequence of i.i.d. random variables having common expectations $E(|X_i|) = m_i = m < \infty$ and a finite variance σ^2.

Let $\overline{X}_n = S_n/n$ be an arithmetic mean where $S_n = \sum_{i=1}^{n} X_i$. The CLT states that

$$U_n = \frac{\overline{X}_n - m}{\sigma/\sqrt{n}} \xrightarrow{D} N(x; 0, 1) \tag{1.6.11}$$

The sequence of random variables U_n is, thus, convergent (according to the distribution) to the standard normal distribution.

1.7 Pseudorandom Number Generation

A *random number generator* (RNG) is a "device" that generates a sequence of numbers that cannot be reasonably predicted better than by a random chance. Various applications of randomness have led to a development of methods for generating random data, of which some have existed since ancient times, for example pure "mechanical" methods like rolling of a die, coin flipping, shuffling of playing cards. True *hardware random-number generators* generate genuinely random numbers. *Pseudorandom number generator* (PRNG) is an algorithm for generating a sequence of numbers whose properties approximate the properties of sequences of random numbers. Careful mathematical analysis is required to have any confidence that PRNG generates numbers that are sufficiently close to random to suit the intended use. PRNGs are central in computer simulations.

An early computer-based PRNG known as the *middle-square method* was suggested by John von Neumann in 1946. This algorithm takes any number, squares it, removes the middle digits of the resulting number as the "random number", then uses that number as the seed for the next iteration and so on. A problem with this method is that all sequences eventually repeat them-selves, some very quickly. In the second half of the twenties century, the standard class of algorithms used for PRNGs were *Linear Congruential Generators* (LCGs). The LCG uses the *recurrence* relation

$$X_{i+1} = (aX_i + b) \bmod M \quad i = 1, 2, \ldots \quad a, b < M \tag{1.7.1}$$

to generate numbers, where a, b and M are large integers, X_0 is a "seed" or start value, X_{i+1} is the next in X_i in a series of pseudorandom numbers (*mod* stands for an operation called *modulus* that finds the remainder after division of one number by another). LCGs are extremely sensitive to the choice of the parameters. A major advance in the construction of pseudorandom generators was the introduction of techniques based on linear recurrences on the two-element field and such generators are related to *linear feedback shift registers*. The invention of the *Mersenne Twister*, in particular, avoided many of the problems with earlier generators.

Most computer programming languages, including R, developed functions or library routines that provide random number generators. They are often designed to provide a random byte or word, or a floating-point number **uniformly distributed** between 0 and 1. These random numbers are then transformed as needed to simulate the different probability distributions. One well known such method is the **inverse transform method** which can be formulated by the following theorem.

Theorem 1.7.1 *Let* F *denotes any cumulative distribution function (cdf). Let*

$$F^{-1}(y) = \min\{x : F(x) \geq y\}, \ y \in [0, \ 1] \tag{1.7.2}$$

denotes the generalized inverse of F. *Define*

$$X = F^{-1}(U) \tag{1.7.3}$$

where U *has the continuous uniform distribution over the interval* (0, 1). *Then* X *is distributed as* F, *that is,* $P(X \leq x) = F(x), \ x \in R$.

The inverse transform method can be used in practice as long as we are able to get an explicit formula for $F^{-1}(y)$ *in closed form. We illustrate this method with example.*

Example 1.7.1 Suppose we have a random variable $U \sim \text{unif}(0, 1)$ and a cdf:

$$F(x) = 1 - e^{-\lambda x} \ \ x \geq 0, \lambda > 0 \tag{1.7.4}$$

Solving the equation $y = 1 - e^{-\lambda x}$ for x in terms of $y \in (0, \ 1)$ yields:

$$x = F^{-1}(y) = -\frac{\ln(1 - y)}{\lambda} \tag{1.7.5}$$

This yields $X = (-1/\lambda)ln(1 - U)$. But from $U \sim \text{unif}(0, 1)$, it follows that $1 - U \sim \text{unif}(0, 1)$ and, thus, we can simplify the algorithm by replacing $1 - U$ by U obtaining the following

Algorithm for generating an exponential random number at rate λ

1. Generate $U \sim \text{unif}(0, 1)$
2. Set $X = (-1/\lambda)ln(U)$

We describe now a simple method for generating random numbers from a standard normal distribution $N(0, 1)$ based on the central limit theorem.

Let's consider the mean

$$\overline{X}_{12} = \frac{X_1 + \ldots + X_{12}}{12} = \frac{S_{12}}{12} \tag{1.7.6}$$

of 12 random variables, each of which has a uniform distribution on the interval (0, 1), e.g. $X_i \sim \text{unif}(0, 1)$. The expectation $E(X_i) = 1/2$ and the variance $D^2(X_i) = 1/12$.

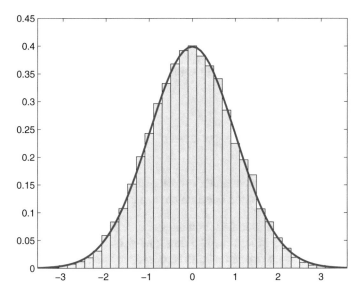

Fig. 1.14 Histogram of standardized sum of 12 random variables from the uniform distribution for 10,000 repetitions. The plot of normal probability density function is also superimposed

From the central limit theorem, a standardized mean:

$$Z = \frac{\overline{X}_{12} - \frac{1}{2}}{\sqrt{1/12}/\sqrt{12}} = S_{12} - 6 \tag{1.7.7}$$

follows a distribution close to the standard normal distribution $N(0, 1)$ which gives a "recipe" for generating random numbers from this distribution.

Figure 1.14 presents the histogram of distribution of such standardized sum of 12 uniformly distributed random variables for 10,000 repetitions of random experiment described by the Formula 1.7.7.

1.8 Computer Simulations of Random Events

A simulation is an approximate imitation of a certain operation in some process or system. The act of simulating requires a simulation model to be developed. This model is a well-defined description of a simulated operation and represents its key characteristics, such as behavior, functions and abstract or physical properties. Often, computer experiments are used to study simulation models.

In this section, we work through examples to show how to apply simulation methods to probability problems and model random events. Computer simulations are available to everyone - in particular R is a really powerful tool for their implementation (the elementary introduction to R is in Appendix B). Consider for example the simulation of rolling a die. The R "runif( )" function generates pseudorandom numbers from uniform distribution on the interval [0, 1]. Let us divide the interval

[0, 1] into six equal parts and assign to each of them one elementary event: for the interval [0, 1/6]—"1 dot", for [1/6, 2/6]—"2 dots", and so on. Sampling a number from interval [0, 1/6] using the mentioned "runif( )" function is as likely as for the remaining five intervals. The code for simulation of rolling a die has the following form.

```
**R**

number_dots = floor(6*runif(1)+1)

***
```

Since we have "forced" a computer to "roll a die", we now show how to design a computer simulation to answer a question about a chance to win in a certain game.

Example 1.8.1 Let's consider the following game. A player pays 2 $ and rolls a die 6 times. If a player gets at least two "6 dots", he wins 6 $, otherwise he loses 2 $ he paid. We are interested in estimation of a chance to win more money than we've paid in 60 such games.

Note that in order to win more than we've paid, we must win more than 12 games in 60 games (at 12 we break even). To answer the question about a chance of financial success in 60 games, we should count the number of simulations in which we win more than 12 games of 60 in a large number of simulations, for example 1,000,000. According to Bernoulli LLN, the resulting frequency is close to the probability of success.

Below, we present R code that "plays" 1,000,000 times, each time 60 games, and, then, returns a percentage of those games with at least 13 wins.

R

```
cnt = 0   # counter
for (r in 1:1000000)
{
cnt60 = 0
    for (i in 1:60)
        {k = 0
        for (j in 1:6) {if (runif(1) > 5/6) k = k + 1}
        if (k >= 2)   cnt60 = cnt60 + 1 }
if (cnt60 > 12) cnt = cnt + 1}
chance = cnt / 1000000
```

***

With the above presented simulation, the estimated chance to win is approximately **0.898606**.

We can also calculate the above probability analytically using the formulas from the Bernoulli trials. The probability of success in a single game is:

$$P(S_6 \geq 2) = 1 - P(S_6 < 2) = 1 - P_{6,1} - P_{6,0}$$

$$= 1 - \binom{6}{1}\left(\frac{1}{6}\right)^1\left(\frac{5}{6}\right)^5 - \binom{6}{0}\left(\frac{1}{6}\right)^0\left(\frac{5}{6}\right)^6 \approx 0.2632 \quad (1.8.1)$$

S_6 stands for the number of successes in 6 rolls. The probability of obtaining more than 12 successes in 60 games is:

$$P(S_{60} > 12) = 1 - P(S_{60} \le 12) = 1 - \left[P_{60,0} + P_{60,1} + \ldots + P_{60,12} \right]$$

$$= 1 - \sum_{i=0}^{12} \binom{60}{i} (0.2632)^i \cdot (1 - 0.2632)^{60-i} \approx 0.8989 \quad (1.8.2)$$

The above, precise calculations are time-consuming and not easy to calculate even with the help of a calculator, arguing in favour of simulation.

Computer simulations can also be used to investigate the distribution of random variables. Such simulations rely on multiple repetitions of random experiment and taking the resulting empirical distribution as an approximation of the probability distribution of random variable under consideration.

Example 1.8.2 Let's consider sampling without replacement n times from exponential distribution with parameter $\lambda = 2$ using "rexp" R function (see R code of simulation below). Let's calculate the empirical means of such generated random variables $X_i, i = 1, \ldots, n$:

$$M_n = \frac{S_n}{n} = \frac{\sum_{i=1}^{n} X_i}{n} \quad (1.8.3)$$

and, then, make a plot of a sequence of increasing means:

$$M_1, M_2, \ldots, M_n \quad (1.8.4)$$

to illustrate their convergence. R code below generates such a sequence from $n = 1000$ element sample.

```
**R**
nmax; = 1000
n = (1:nmax)
lambda = 2
X = rexp(nmax,rate = lambda)
S = cumsum(X)
M = S/n
Plot(n,M,type="l")
***
```

Figure 1.15 shows the plot produced by the above simulation. As you can see, according to the LLN as the sample size increases, the mean of many independent realizations of a random variable tends to its expected value, here equal $1/\lambda = 1/2$.

Example 1.8.3
Let us now simulate the convergence of sequence of distributions as described by

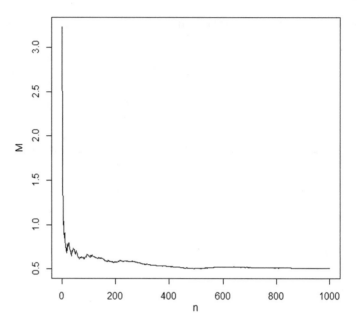

Fig. 1.15 A plot of sequence of increasing means for 1000 repetitions of random experiment

CLT. Consider the sequence of independent random variables from exponential distribution with parameter $\lambda = 2$ and plot the histogram of empirical distribution of their sums for 10,000 repetitions (Fig. 1.16). The appropriate R code is given below.

```
**R**
m = 10000
n = 100
lambda = 2
S = replicate(m,sum(rexp(n,rate = lambda)))
hist(S,prob="TRUE")
curve(dnorm(x,mean = n/lambda,sd = sqrt(n)/lambda),
col = "blue",add = TRUE)
***
```

The "replicate" function enables multiple execution of an operation specified as an argument—here the generation of random numbers from exponential distribution and, then, calculation of their sum. From the obtained histogram, it follows that the fitting to the normal density curve is good. Assessing the goodness of fit formally requires a statistical test, which will be discussed later.

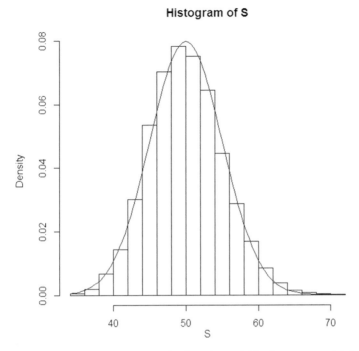

Fig. 1.16 Histogram of distribution of sums of 100 random variables from exponential distribution for 10,000 repetitions of random experiment with normal density curve superimposed

Bibliography

Billingsley, P.: Probability and Measure, 3rd edn. Wiley, New York (1995)

Hodges Jr., J.L., Lehmann, E.L.: Basic Concepts of Probability and Statistics, 2nd edn. Society for Industrial and Applied Mathematics, Philadelphia (2005)

Papoulis, A., Pillai, S.U.: Probability, Random Variables and Stochastic Processes, 4th edn. McGraw-Hill, New York (2002)

R Core Team: R language definition. http://cran.r-project.org/doc/manuals/r-release/R-lang.pdf (2019). Accessed 16 Dec 2019

Rudin, W.: Principles of Mathematical Analysis, 3rd edn. McGraw-Hill, New York (1976)

Soong, T.T.: Fundamentals of Probability and Statistics for Engineers, 1st edn. Wiley, Chichester (2004)

Chapter 2
Descriptive and Inferential Statistics

2.1 Preliminary Concepts

Merriam-Webster dictionary defines *statistics* as "a branch of mathematics dealing with the collection, analysis, interpretation, and presentation of masses of numerical data" (www.merriam-webster.com *retrieved 2016-05-28*).

Statistical concepts and methods are not only useful but indeed often indispensable in understanding the world around us. They provide ways of gaining new insights into the behavior of many phenomena that you will encounter in your chosen field of specialization in engineering or science. The discipline of statistics teaches us how to make intelligent judgments and informed decisions in the presence of *uncertainty* and *variation*. Without uncertainty or variation, there would be little need for statistical methods or statisticians. If every component of a particular type had exactly the same length, if all resistors produced by a certain manufacturer had the same resistance value, and so on, then a single observation would reveal all desired information.

Example 2.1.1 below illustrates the manifestation of variation.

Example 2.1.1 Automatic salt packaging into 1 kg bags in a factory

Complaints from customers who purchased 1 kg bags of salt produced in a certain factory indicate underweighting. To check the legitimacy of complaints, the quality control in a factory starts a *statistical study*.

The 25 salt bags have been randomly sampled from a whole production of factory (called here a **population**) and weighted on an independent weighting machine.

The results (in grams) which make up our **Sample No. 1** are given below.
*

   1000.33, 1004.97, 998.98, 1000.85, 1001.42, 1001.68, 999.58, 1001.16,
1001.79, 997.64, 1001.59, 1000.56, 1003.26, 996.25, 995.83, 999.56,
1002.08, 998.89, 998.09, 1004.42, 1002.14, 998.01, 1002.79, 999.56,
1001.72

© Springer Nature Switzerland AG 2020

K. Stapor, *Introduction to Probabilistic and Statistical Methods with Examples in R*,
Intelligent Systems Reference Library 176,
https://doi.org/10.1007/978-3-030-45799-0_2

*

Before we can understand what a particular sample can tell us about the population, we should first understand the **uncertainty** associated with taking a sample from a given population. This is why we have been studying probability before statistics.

We would like to emphasize the differences between approaches to a problem in probability and statistics. In a probability problem, properties of the population under study are assumed known (e.g., some specified distribution), and questions regarding a sample taken from the population are posed and answered. In an inferential statistics problem, characteristics of a sample are available, and this information enables to draw conclusions about the population. The relationship between the two disciplines can be summarized by saying that probability reasons from the population to the sample (*deductive reasoning*), whereas inferential statistics reasons from the sample to the population (*inductive reasoning*).

In this chapter, we will present the four main stages in doing *statistical study* like the one above:

1. designing a study,
2. collecting the data,
3. obtaining descriptive statistics,
4. performing some inferential statistics.

As a result of performing such a four-step study, we'll be able to draw conclusions like "***whether the mean weight of salt bags is conforming to manufacturing specifications or not***" based on different methods of statistical inference.

The question about the mean weight of salt bags produced in a factory (Example 2.1.1) is, in fact, a question about an expected value $E(X)$ of distribution of salt bags weights in a whole population of manufactured salt bags. The answer can be obtained by methods of statistical inference in three different ways: by *point* or *interval estimation*, or by *testing statistical hypothesis* which will be presented throughout this chapter.

Population, Sample

A statistical study like the one above will typically focus on a well-defined collection of objects constituting *a population* of interest. In the study above, the population might consist of all salt bags produced during a specified period. When desired information is available for all objects in the population, we have what is called *a census*. Constraints on time, money, and other scarce resources usually make a census impractical or infeasible. Instead, a subset of the population, *a sample*, is selected in some prescribed manner (as in our Example 2.1.1, the set of 25 salt bags sampled from a particular production). A *sampling frame*, that is, a listing of the individuals or objects to be sampled is either available to an investigator or else can be constructed.

Types of Data

We are usually interested only in certain characteristics of the objects in a population: the weight of salt bags, the number of flaws on the surface of each casing, the gender of an engineering graduate, and so on. A *variable* (or *feature*) is any characteristic whose value may change from one object to another in the population. Generally, we are dealing with four types of data.

Nominal (*categorical*) data are numbers that are used only as names/labels for categories. If, for example, you had two groups, you could label them "1", "2", but it would mean just the same as if they were entitled "A", "B". You cannot do any statistics or arithmetic on nominal data.

Ordinal (*rank*) data can be ordered in terms of some property such as size, length, etc. However, successive points on the scale are not necessarily spaced equally apart. To rank the swimmers in terms of whether they finish first, second, and so on, is to measure them on an ordinal scale: all that you know for certain is the order of finishing—the difference between first and second placed swimmers may be quite different from the difference between those placed second and third.

Interval data are measured on a scale on which measurements are spaced at equal intervals, but on which there is no true zero point. The classic example is the Centigrade temperature scale. On the Centigrade scale, a rise of "one degree" represents an increase in heat by the same amount, whether one is talking about the shift from 2 to 3°, or from 25 to 26°. However, because the zero point on the scale is arbitrary, one cannot make any statements like "4 degrees is twice 2 degrees".

Ratio scale is the same as the interval scale, except that there is a true zero on the scale, representing an absence of the measured thing. Measurements of weight, height, length, time are all measurements on ratio scales. With this kind of scale, the intervals are equally spaced and we can make meaningful statements about the ratios of quantities measured.

Data results from making observations either on a single variable or simultaneously on two or more variables. An *univariate* data set consists of observations on a single variable, while *multivariate* data arises when observations are made on more than one variable.

Descriptive Versus Inferential Statistics
A statistician who has collected data may wish simply to summarize and describe important features of the data. This entails using methods from ***descriptive statistics***. Some of these methods are graphical in nature. The construction of histograms, boxplots, and scatter plots are primary examples. Other descriptive methods involve calculation of numerical summary measures, such as means, standard deviations, and correlation coefficients. Just look at the data in the Sample Nr. 1 in Example 2.1.1. Without any organization, it is difficult to get a sense of the data's most prominent features, what a typical value might be, whether values are highly concentrated about a typical value or rather dispersed, what fraction of the values are less than 980 g, and so on.

Having obtained a sample from a population, a statistician would frequently like to use sample information to draw some type of conclusion about the population, here about a mean weight of salt bags. Techniques for generalizing from a sample to a population are gathered within the branch of our discipline called ***inferential statistics*** which may be divided into two major areas:

- *Estimation*, i.e. approximating the values of parameters or functional form of population distribution.
- *Tests of hypothesis* about parameter values or functional form of population distribution.

Any statistical inference requires some *assumptions*, i.e. a *statistical model* concerning the generation of the observed data. In general, statisticians distinguish between two such models:

- *Parametric*: the probability distributions describing the data-generation process are assumed to be fully described by a family of probability distributions involving only a finite number of unknown parameters. For example, one may assume that the distribution of population values is normal, with unknown mean and variance.
- *Non-parametric*: the assumptions made about the process generating the data do not rely on data from any particular parametric family of probability distributions.

Types of Statistical Studies

According to the purpose of the analysis, two main types of studies are conducted. In *exploratory data analysis*, the purpose is to investigate the data to see what patterns can be seen. In *confirmatory data analysis*, a pattern has been hypothesized before the study, and the purpose of the study is to confirm or disconfirm the hypothesis.

According to how the data are collected, there are three basic types of statistical studies. An *observational study* measures the characteristics of a population by studying individuals in a sample, but does not attempt to manipulate or influence the variables of interest. The easiest examples of observational studies are surveys. Another example is the statistical study in Example 2.1.1.

In contrast, *designed experiments* explicitly do attempt to influence results. For example, imagine that statistician is interested in comparing two methods for teaching reading. He randomly assigns half the schools in their sample to one method and the other half to the other method. At the end of the school year, he analyzes reading scores of the children in the schools. This is an *experiment*: statistician deliberately decides which students receive each teaching method.

Since observational studies don't control any variables, the results can only be associations. Because variables are controlled in a designed experiment, we can have conclusions of causation.

A *retrospective studies* investigate a phenomenon that has occurred in the past. Such studies most often involve secondary data collection, based upon data available from previous studies or databases. For example, a retrospective study would be needed to examine the relationship between levels of unemployment and street crime in some place over the past 100 years.

Collecting Data, Sampling Procedures

Statistics deals not only with the organization and analysis of data once it has been collected, but also with the development of techniques for collecting the data. If data is not properly collected, an investigator may not be able to answer the questions under consideration with a reasonable degree of confidence.

When data collection entails selecting individuals or objects from a frame, the simplest method for ensuring a representative selection is to take a *simple random sample*. This is one for which any particular subset of the specified size (e.g., a sample of size 100) has the same chance of being selected. For example, if the frame consists of 1,000,000 serial numbers, the numbers 1, 2, ..., up to 1,000,000 could be placed on identical slips of paper. After placing these slips in a box and thoroughly mixing, slips could be drawn one by one until the requisite sample size has been obtained. Alternatively (and much to be preferred), a table of random numbers or a computer's random number generator could be employed.

Sometimes, alternative sampling methods like stratified, cluster, systematic and other ones can be used to make the selection process easier, to obtain extra information, or to increase the degree of confidence in conclusions.

2.2 Descriptive Statistics

The next step after the completion of data collection is to organize or represent the data into a meaningful form, so that a trend, if any, emerging out of the data can be seen easily. For this purpose, the methods of descriptive statistics can be used. There are two general subject areas of representing a data set:

1. using tabular forms and visual techniques like frequency tables, histograms,
2. using some numerical summary measures describing location of central tendency, measures of data variability, asymmetry or kurtosis.

2.2.1 Tabular and Pictorial Methods

Tabular Methods
One of the common methods for organizing data is to construct frequency distribution. A *frequency distribution* is a table or graph that displays the frequency of various outcomes/observations in a sample: each entry in the *frequency table* contains the frequency, or, count of the occurrences of values within a particular group or interval, and in this way, the table summarizes the distribution of values in the sample.

Frequency distribution tables (or shortly *frequency tables*) can be used for both discrete and continuous variables. They can be constructed by grouping observations, i.e. the realizations $x_1, x_2, \ldots, x_n$ of some random variable X, which stands as a "model" of our data, into certain *classes* that can be:

1. *Points (variants)* x_i, $i = 1, \ldots, k$, where k is the number of different values of discrete random variable X.

2. *Intervals* $[x_{0i}, x_{1i})$, $i = 1, ..., k$, where k is the accepted number of intervals, if X is continuous random variable or the number of variants exceeds the assumed maximum number (generally 6–14 variants are acceptable).

Grouping observations into a certain class i relies on counting the number n_i of times the observation occurs in the data. We construct a table with two columns. In the first column, we write down all the variants, or, intervals of the data values in ascending order of magnitude. To complete the second column, we calculate the *relative frequency* of each class i, $i = 1, \ldots, k$:

$$w_i = \frac{n_i}{n} \qquad (2.2.1.1)$$

where $n = \sum_{i=1}^{k} n_i$ is a sample size, and k is the number of classes.

Before constructing a frequency table, one should have an idea about the number k of classes. Too small, will not allow revealing the shape distribution of the feature in the sample, and too much, will result in excessive detail of the description. It is often assumed that the number k is proportional to the square root of the number of measurements n. Then, we should decide the width of the classes. Calculating the range $(x_{max} - x_{min})$ of the data, where x_{min}, x_{max} are the minimum and the maximum values in a sample, we can determine the *class width*:

$$h = \frac{x_{max} - x_{min}}{k} \qquad (2.2.1.2)$$

Generally, the class width is the same for all classes. Note, that equal class intervals are preferred in frequency distribution, while unequal class interval may be necessary in certain situations to avoid empty classes. Table 2.1 shows the frequency distribution table for the Sample Nr. 1 (Example 2.1.1).

Frequency distribution allows us to have a glance at the entire data conveniently. It shows, whether the observations are high or low, and also, whether they are concentrated in one area or spread out across the entire scale.

Let us denote by r the number of sample elements which take values lower than a fixed number x:

Table 2.1 Frequency distribution table for Sample Nr. 1 (Example 2.1.1)

	From	To	Frequency	Rel. freq.	Cum. freq.
1	995.827	997.352	2	0.08	0.08
2	997.352	998.876	3	0.12	0.20
3	998.876	1000.401	6	0.24	0.44
4	1000.401	1001.925	8	0.32	0.76
5	1001.925	1003.450	4	0.16	0.92
6	1003.450	1004.974	2	0.08	1.00

$$r = \sum_{i=1}^{n} 1(x_i < x) \qquad (2.2.1.3)$$

where

$$1(\text{arg}) = \begin{cases} 1 & \text{arg} = T \\ 0 & \text{arg} = F \end{cases} \qquad (2.2.1.4)$$

T, F—logical values "true", "false".
   The following function:

$$\hat{F}_n(x) = \frac{r}{n} = \frac{\sum_{i=1}^{n} 1(x_i < x)}{n} = \begin{cases} 0 & x < x_1 \\ \sum_{s=1}^{i} w_s\ x_{0i} \le x \le x_{1i} \\ 1 & x > x_k \end{cases} \qquad (2.2.1.5)$$

is the *empirical cumulative distribution function*. This function is an empirical equivalent of an unknown cumulative distribution function $F(x)$ of random variable X (i.e. in population). This correspondence is expressed by *Gliwienko's theorem*, which tells of the uniform convergence with probability 1 of the sequence of random variables $\hat{F}_n(x)$ to $F(x)$. This means that, if the sample is sufficiently large, the empirical distribution differs little from the theoretical, unknown distribution.

Pictorial Methods

The information provided by a frequency distribution table is easier to grasp if presented graphically. The most popular is a *histogram* which, for interval frequency table, is a set of rectangles, whose width is equal to the length of class interval while their heights are equal to the relative frequency divided by the class width (see the exampled histogram in Fig. 2.2). In the case of point frequency tables, the rectangles in the histogram are reduced to the vertical bars at the succeeding values of random variable. As the size of the sample becomes larger, the resulting limit histogram would be the plot of unknown probability distribution of X.

   A distribution is said to be *symmetric* if it can be folded along a vertical axis so that the two sides coincide. A distribution that lacks symmetry with respect to a vertical axis is said to be *skewed*. The distribution illustrated in Fig. 2.2 is said to be slightly skewed to the left since it has a little left tail and a shorter right tail.

   Another display that is helpful for reflecting properties of a sample is the *box-and-whisker plot* (see the example in the bottom of Fig. 2.2). This plot encloses the interquartile range i (for explanations see next subsection) of the data in a box that has the median displayed within. The interquartile range i has as its extremes the 75th percentile (upper quartile) and the 25th percentile (lower quartile). For reasonably large samples, the display shows center of location, variability and the degree of symmetry: the position of the median symbol relative to the two edges conveys information about skewness in the middle 50% of the data. In addition to the box, "whiskers" extend, showing extreme observations in the sample. A boxplot can be embellished to indicate explicitly the presence of outliers. Any observation farther than *1.5i* from the closest quartile is an *outlier*. An outlier is *extreme* if it is more than *3i* from the nearest quartile, and it is *mild* otherwise.

2.2.2   Summary Measures of Frequency Distribution

Visual summaries of data are the excellent tools for obtaining preliminary impressions and insights. More formal data analysis often requires the calculation and interpretation of numerical summary measures. That is, from the data we try to extract several summarizing numbers that might serve to characterize the data set and convey some of its salient features.

In general, four groups of summary measures (statistics) are distinguished:

- measures of location of a data center,
- measures of variability,
- measures of symmetry,
- measures of kurtosis.

Measures of Location
Location measures in a sample are designed to give some quantitative measure of the location (center) of a sample. A common measure is the arithmetic mean:

$$\bar{x} = \frac{1}{n} \sum_{i=1}^{n} x_i \tag{2.2.2.1}$$

It is called the *sample mean* and is the point which "balances" the system of "weights" corresponding to the values of elements in a sample. The sample mean can be greatly affected by the presence of even a single outlier (unusually large or small observation).

Quantiles (empirical) are the other measures of central tendency that are uninfluenced by outliers. Formally, *quantile of order p* ($0 < p < 1$) is the lowest value x_p in a sample, for which the empirical cumulative distribution function satisfies the condition:

$$\hat{F}_n(x_p) \geq p \tag{2.2.2.2}$$

Quantiles can be computed exactly from data available, or, based on interpolation expressions built on frequency distribution tables. For example, given that the observations $x_1 \leq x_2 \leq \ldots \leq x_n$ in a sample are arranged in increasing order of magnitude, the *sample median*, which is a quantile of order $p = 1/2$, can be computed as:

$$me = q_2 = \begin{cases} x_{(n+1)/2} & n \text{ is odd} \\ \frac{1}{2}(x_{n/2} + x_{n/2+1}) & n \text{ is even} \end{cases} \tag{2.2.2.3}$$

Sample median is an example of *quartiles* which divide the sample into four equal parts and *me* is the *second quartile* q_2. The observations above the *third quartile* q_3 constitute the upper quarter of the sample, and the *first quartile* q_1 separates the lower quarter from the upper three-quarters. The quartile q_1 can be obtained as the one, which position in the increasingly ordered sample is $(n + 1)/4$ (with rounding up). Similarly, the quartile q_3 is calculated as the one, which position in the increasingly ordered sample is $3(n + 1)/4$ (with rounding down).

A sample can be even more finely divided using *percentiles*: the 99th percentile separates the highest 1% from the bottom 99%, and so on.

The *mode* is the value which occurs most frequently in a sample.

Measures of Variability

There are many measures of variability. The simplest one is the *sample range*:

$$r = x_{max} - x_{min} \tag{2.2.2.4}$$

The similar measure that is "resistant" to the presence of outliers is the *interquartile range*:

$$i = q_3 - q_1 \tag{2.2.2.5}$$

The most often used measure is the *sample variance*:

$$s^2 = \frac{1}{n} \sum_{i=1}^{n} (x_i - \bar{x})^2 \tag{2.2.2.6}$$

which measures the average squared deviation from an arithmetic mean $\bar{x}$, and the *sample standard deviation*, the positive square root of s^2:

$$s = \sqrt{s^2} \tag{2.2.2.7}$$

Large variability in a data produces relatively large $(x_i - \bar{x})^2$, and, thus a large sample variance.

Coefficient of Variation:

$$v = \frac{s}{\bar{x}} \tag{2.2.2.8}$$

also known as *relative standard deviation*, is a standardized measure of dispersion of a frequency distribution, often used to compare the populations.

Measures of Symmetry

Skewness is a measure of the asymmetry of the probability distribution. The already introduced coefficient *As* of skewness of random variable (also known as *Pearson's moment*) can be estimated from a sample resulting in *sample skewness*:

$$as = \frac{\frac{1}{n} \sum_{i=1}^{n} (x_i - \bar{x})^3}{s^3} \tag{2.2.2.9}$$

Values equal to zero are reserved for symmetrical distribution, positive for right asymmetry, negative for left asymmetry. Skewness is positive if the tail on the right side of the distribution is longer, or, fatter than the tail on the left side.

Measures of Kurtosis

Kurtosis is a measure of the "tailedness" of the probability distribution of a real-valued random variable, and there are different ways of quantifying it. The standard measure of kurtosis, originating with Karl Pearson, is based on a scaled version of the fourth moment of the data and can be estimated from a sample resulting in *sample kurtosis*:

$$krt = \frac{\frac{1}{n} \sum_{i=1}^{n} (x_i - \bar{x})^4}{s^4} \tag{2.2.2.10}$$

The kurtosis of any univariate normal distribution is equal to 3. Distributions with kurtosis less than 3 are said to be *platokurtic* and it means the distribution produces fewer and less extreme outliers than does the normal distribution (for example the uniform distribution, which does not produce outliers). Distributions with kurtosis greater than 3 are said to be *leptokurtic* (for example Laplace distribution).

It is also common practice to use an adjusted version of Pearson's kurtosis, the *excess kurtosis*, which is the kurtosis minus 3, to provide the comparison to the normal distribution.

Example 2.2.2.1 This example illustrates the described methods of descriptive statistics to obtain a meaningful form of data from Sample No. 1 (Example 2.1.1).

Table 2.1 is a frequency distribution table constructed based on $k = 6$ intervals. By grouping the observation from Sample No. 1 to appropriate class-intervals, we obtain their frequencies. In the last two columns, there are the relative frequencies and empirical cumulative distribution function, respectively.

Figure 2.1 shows the histogram based on the constructed frequency distribution table. The plot of probability density $N(x; 1000, 3)$ is superimposed, to visually assess the fit of empirical distribution. The theoretical assessment of the goodness of fit requires the methods of statistical inference which will be presented in the next subsection.

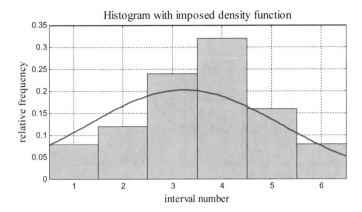

Fig. 2.1 Histogram based on distribution Table 2.1 from Sample Nr. 1 with a plot of the normal probability density superimposed

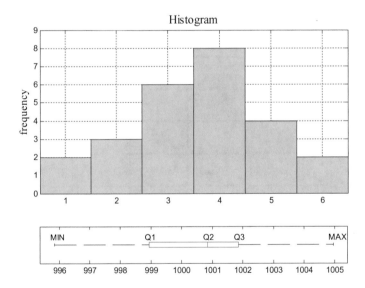

Fig. 2.2 Histogram for Sample Nr. 1 with box-and-whisker plot in the bottom part

Figure 2.3 presents a plot of empirical cumulative distribution function based on the constructed frequency distribution table. The individual points were additionally connected by a dashed line which results from the assumption of uniform distribution within the interval, resulting in a linear increase in frequency. The distribution function is a step function with constant "jumps".

Figure 2.4 shows a plot of empirical cumulative distribution function based on raw sample data. The corresponding calculations are shown in Table 2.2. This is also a step function, with constant "jumps" equal to $1/n$ (where n is the sample size)

Empirical cumulative distribution based on frequency distribution

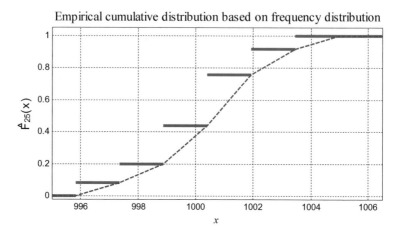

Fig. 2.3 The empirical cumulative distribution function based on distribution Table 2.1 for Sample Nr. 1

Empirical cumulative distribution based on raw data

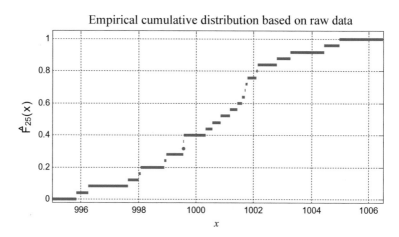

Fig. 2.4 The empirical cumulative distribution function based on raw data from Sample Nr. 1

in the observed points x_i. The minimum and maximum values are: $x_{min} = 995.83$, $x_{max} = 1004.97$ which gives the value of the range $r = 9.14$. The sample mean, is $\bar{x} = 1000.53$. In order to calculate the quartiles, we order sample data as shown in Table 2.3. The position of the q_1 quartile is $k = \frac{25+1}{4} \approx 7$, so the 7th element from the ordered sample is the quartile $q_1 = 998.98$. The q_3 quartile occupies a position $k = \frac{3(25+1)}{4} \approx 19$, hence its value is $q_3 = 1001.80$. The median takes a position $k = \frac{25+1}{2} = 13$, hence its value $q_2 = 1000.85$. The interquartile range is $i = q_3 - q_1 = 2.82$. This range includes the central 50% of the observations in the sample.

Table 2.2 Calculations of empirical cumulative distribution function based on raw data from Sample Nr. 1

x	$\hat{F}_{25}(x)$	x	$\hat{F}_{25}(x)$
995.83	0.04	1001.16	0.56
996.25	0.08	1001.42	0.60
997.64	0.12	1001.59	0.64
998.01	0.16	1001.68	0.68
998.09	0.20	1001.72	0.72
998.89	0.24	1001.80	0.76
998.98	0.28	1002.08	0.80
999.56	0.32	1002.14	0.84
999.56	0.36	1002.80	0.88
999.58	0.40	1003.26	0.92
1000.33	0.44	1004.42	0.96
1000.56	0.48	1004.97	1.00
1000.85	0.52		

Table 2.3 The observations from Sample Nr. 1 in non-descending order

No.	x	No.	x
1	995.83	14	1001.16
2	996.25	15	1001.42
3	997.64	16	1001.59
4	998.01	17	1001.68
5	998.09	18	1001.72
6	998.89	19	$q_3 = \mathbf{1001.80}$
7	$q_1 = \mathbf{998.98}$	20	1002.08
8	999.56	21	1002.14
9	999.56	22	1002.80
10	999.58	23	1003.26
11	1000.33	24	1004.42
12	1000.56	25	1004.97
13	$q_2 = \mathbf{1000.85}$		

The obtained 3 quartiles are shown in bold

The calculated sample and interquartile ranges characterize in some way the variability of weights in the Sample Nr. 1. The standard deviation is $s = \sqrt{5.13} = 2.27$ and it represents *23%* of the sample mean (the coefficient of variation is $v = \frac{\bar{x}}{s} = 0.23$). The sample skewness $as = -0.17$ confirms the already observed slight, left asymmetry of the empirical distribution. The sample kurtosis is $krt = 2.39 < 3$, so fewer outliers in comparison with normal distribution.

In the bottom part of Fig. 2.2, under the histogram, there is a *box-and-whisker plot* for Sample Nr. 1. The median is slightly shifted to the right in relation to the interquartile range, which indicates a slight left asymmetry (i.e. more elements in the sample are concentrated to the left of the median).

2.3 Fundamental Sampling Distributions

We begin this section by discussing more formally the notions of populations and samples which were already introduced in the previous sections. However, much more needs to be discussed about them here, particularly in the context of random variables. The totality of observations with which we are concerned (finite or infinite) constitutes, what we call a *population*. Groups of people, animals, or all possible outcomes from some complicated engineering system are the examples. Each *observation* in a population is a value of some random variable X having some probability distribution $f_X(x)$ (or $p_X(x)$ if X is of discrete type). For example, if one is inspecting items coming off an assembly line for defects, then each observation x in the population might be a value 0 or 1 of the Bernoulli random variable X with probability distribution

$$b(x; 1, p) = p^x q^{1-x} \quad x = 0, 1 \tag{2.3.1}$$

where 0 and 1 indicate a nondefective or defective item, respectively. It is assumed that p, as well as q, the probabilities of any item being defective or nondefective, remains constant from trial to trial. When we refer hereafter to a "*binomial population*", a "*normal population*" or, in general, the "*population $f_X(x)$*", we shall mean a population whose observations are values of a random variable having a binomial distribution, a normal distribution, or the probability distribution $f_X(x)$. Hence, the mean and variance of a random variable or probability distribution are also referred to as the mean and variance of the corresponding population.

In the field of statistical inference, the statistician is interested in arriving at conclusions concerning a population when it is impossible, or, impractical to observe the entire set of observations that make up the population. For example, in attempting to determine the average length of life of a certain brand of light bulb, it would be impossible to test all such bulbs if we are to have any left to sell. This brings us to consider the notion of *sampling*.

A *sample* is a subset of a population. If our inferences from the sample to the population are to be valid, we must obtain samples that are *representative* of the population. We are often tempted to choose a sample by selecting the most convenient members of the population. Such a procedure may lead to erroneous inferences concerning the population. To eliminate any possibility of such bias in the sampling procedure, it is desirable to choose a *random sample* in the sense that the observations are made independently, and, at random.

In selecting a random sample of size n from a population $f_X(x)$, represented here by a random variable X, let us define the random variable X_i, $i = 1, 2, \ldots, n$, to represent the ith measurement, or, sample value that we observe. The random variables $X_1, X_2, \ldots, X_n$ will then constitute a *random sample* from the population $f_X(x)$ with numerical values $x_1, x_2, \ldots, x_n$, if the measurements are obtained by repeating the experiment n independent times under essentially the same conditions. Because of the identical conditions under which the elements of the sample are selected, it is reasonable to assume that the n random variables $X_1, X_2, \ldots, X_n$ are independent, and, that each has the same probability distribution $f_X(x)$.

The random variables $X_1, X_2, \ldots, X_n$ satisfying the described conditions:

- $X_1, X_2, \ldots, X_n$ are independent,

- $f_{X_i}(x) = f_X(x)$ for all x, $i = 1, 2, \ldots, n$.   (2.3.2)

are called a *random sample* of size n, and is denoted here as $(X_1, X_2, \ldots, X_n)$.

If X is a random variable of the discrete type with probability mass function $p_X(x)$, then $p_{X_i}(x) = p_X(x)$ for all x, and $i = 1, 2, \ldots, n$.

Our main purpose in selecting random samples is to elicit information about the unknown population parameters. Suppose, for example, that we wish to arrive at a conclusion concerning the mean weight $E(X)$ of salt bags produced by some factory. It would be impossible to check every produced bag in order to compute the mean value of weight representing the population weight. Instead, a random sample is selected to be weighted and based on its realization $(x_1, x_2, \ldots, x_n)$, the mean weight, is then calculated as:

$$\bar{x} = \frac{1}{n} \sum_{i=1}^{n} x_i \tag{2.3.3}$$

The value $\bar{x}$ is then used to make an inference concerning the true $E(X)$. Now, $\bar{x}$ is a function

$$\bar{x} = h(x_1, \ldots, x_n) \tag{2.3.4}$$

of the observed values $x_1, x_2, \ldots, x_n$ in the random sample $(X_1, \ldots, X_n)$. Since many random samples are possible from the same population, we would expect $\bar{x}$ to vary somewhat from sample to sample. That is, $\bar{x}$ is a value of a random variable that we represent by $\overline{X}$:

$$\overline{X} = h(X_1, \ldots, X_n) \tag{2.3.5}$$

Such a random variable is called a *statistic*. Thus, any function of the random variables constituting a random sample is called a *statistic*. In the above example, the chosen statistic h is:

$$\overline{X} = h(X_1, \ldots, X_n) = \frac{1}{n} \sum_{i=1}^{n} X_i \tag{2.3.6}$$

and is commonly called the *sample mean* of population X. The term sample mean is applied to both the statistic $\overline{X}$ and its computed value $\bar{x}$.

Another example of statistic may be:

$$S^2 = \frac{1}{n} \sum_{i=1}^{n} (X_i - \overline{X})^2 \tag{2.3.7}$$

for measuring the population variability and is called the *sample variance* of population X. The statistic $S = \sqrt{S^2}$ is known as *sample standard deviation*.

For the reasons presented in the next section, the following statistic:

$$S_*^2 = \frac{1}{n-1} \sum_{i=1}^{n} (X_i - \overline{X})^2 \tag{2.3.8}$$

is also used as a measure of population variability, especially, when a sample is small.

Since a statistic is a random variable that depends only on the observed sample, it must have a probability distribution. The probability distribution of a statistic is called a *sampling distribution*. The sampling distribution of a statistic depends on the underlying distribution of the population, size of the population, the size of the samples, and the method of choosing the samples.

For the remainder of this section we present several of the more important sampling distributions of frequently used statistics. Applications of these sampling distributions to problems of statistical inference are considered throughout most of the remaining sections.

We assume that the considered populations that are sampled are **normal**.

The following theorem forms a basis for inferencing about a population mean.

Theorem 2.3.1 (sampling distribution of standardized sample mean)

If $\overline{X}$ is the mean of a random sample of size n taken from a population with a normal distribution $N(x; m, \sigma)$, where σ is known, then the statistic:

$$U_1 = \frac{\overline{X} - m}{\sigma} \sqrt{n} \tag{2.3.9}$$

has a standard normal distribution $N(x; 0, 1)$.

A far more important application involves two populations. A scientist, or, engineer is interested in a comparative experiment in which two manufacturing methods, say 1 and 2, are to be compared. The basis for that comparison is $\overline{X}_1 - \overline{X}_2$, the difference in the population means. The following theorem enables statistical inference about such a difference.

Theorem 2.3.2 (sampling distribution of the difference between two sample means)

*If independent samples of size n_1 and n_2 are drawn at random from two normal populations $N(x; m_1, \sigma_1)$ and $N(x; m_2, \sigma_2)$, respectively, in which σ_1, σ_2 **are known**, then, the statistic:*

$$U_2 = \frac{\overline{X}_1 - \overline{X}_2 - (m_1 - m_2)}{\sqrt{\frac{\sigma_1^2}{n_1} + \frac{\sigma_2^2}{n_2}}} \tag{2.3.10}$$

has a standard normal distribution $N(x; 0, 1)$.

In many experimental scenarios knowledge of a standard deviation σ is certainly no more reasonable than knowledge of the population mean. Often, an estimate of σ must be supplied by the same sample information that produced the sample average. As a result, a natural statistic to consider to deal with inferences on sample mean is a T statistic given below, whose distribution formalizes the following theorem.

Theorem 2.3.3 (sampling distribution of studentized sample mean)
If $\overline{X}$ is a mean of a random sample of size n taken from a population with a normal distribution $N(x; m, \sigma)$, where σ is unknown, then the statistic:

$$T = \frac{\overline{X} - m}{S} \sqrt{n - 1} \tag{2.3.11}$$

where S is the sample standard deviation has t-distribution with $(n - 1)$ degrees of freedom.

When we want to compare the unknown expected values m_1, m_2 in two different normal populations with unknown (but equal) variances, we use the following theorems for inference.

Theorem 2.3.4 (sampling distribution of the difference between two sample means)
If independent samples of size n_1 and n_2 are drawn at random from two normal populations $N(x; m_1, \sigma)$ and $N(x; m_2, \sigma)$, respectively, in which $\sigma = \sigma_1 = \sigma_2$, the common variance is unknown, then the statistic:

$$T = \frac{\overline{X}_1 - \overline{X}_2}{\sqrt{\frac{n_1 \cdot S_1^2 + n_2 \cdot S_2^2}{n_1 + n_2 - 2}\left(\frac{1}{n_1} + \frac{1}{n_2}\right)}} \tag{2.3.12}$$

has a t-distribution with $(n_1 + n_2 - 2)$ degrees of freedom.
Here $S_1^2 = \frac{1}{n_1}\sum_{i=1}^{n_1}(X_i - \overline{X}_1)^2, S_2^2 = \frac{1}{n_2}\sum_{i=1}^{n_2}(X_i - \overline{X}_2)^2$ are sample variances.

The above stated theorems are based on sampling distribution of $\overline{X}$ statistic. Sampling distributions of important statistics allow us to learn information about parameters. Usually, the parameters are the counterpart to the statistics in question. If an engineer is interested in the population mean weight of salt bags produced by a certain factory, the sampling distribution of $\overline{X}$ will be exploited once the sample information is gathered. On the other hand, if the variability in salt bag's weights is to be studied, clearly the sampling distribution of S^2 will be used in learning about the parametric counterpart, the population variance σ^2. This formalizes the following theorem.

Theorem 2.3.5 (sampling distribution of sample variance)
If S^2 is a variance of a random sample of size n taken from a normal population having the variance σ^2, then the statistic:

$$\chi^2 = \frac{nS^2}{\sigma^2} \tag{2.3.13}$$

has a chi-squared distribution with $v = n - 1$ degrees of freedom.

While it is of interest to let sample information shed light on two population means, it is often the case that a comparison of variability is equally important, if not more so. The F-distribution finds enormous application in comparing sample variances.

Theorem 2.3.6 (sampling distribution of two sample variances)
If

$$S_{*1}^2 = \frac{1}{n_1 - 1} \sum_{i=1}^{n_1} (X_i - \overline{X}_1)^2 \tag{2.3.14}$$

$$S_{*2}^2 = \frac{1}{n_2 - 1} \sum_{i=1}^{n_2} (X_i - \overline{X}_2)^2 \tag{2.3.15}$$

are the variances of independent random samples of size n_1 and n_2, taken from normal populations with variances σ_1^2 and σ_2^2, respectively, then the statistics:

$$F = \frac{S_{*1}^2/\sigma_1^2}{S_{*2}^2/\sigma_2^2} \tag{2.3.16}$$

has F-distribution with $(n_1 - 1)$ and $(n_2 - 1)$ degrees of freedom.

2.4   Estimation

As was stated in Sect. 2.1, the field *of statistical inference* consists of those methods used to make decisions or to draw conclusions about a population. Statistical inference may be divided into two major areas: *estimation* and *tests of hypotheses*. We treat these two areas separately, dealing with estimation in this section and hypothesis testing in the next.

Statisticians distinguish between the *classical methods* of estimating a population parameter, whereby inferences are based strictly on information obtained from a random sample selected from the population, and the *Bayesian method*, which utilizes prior subjective knowledge about the probability distribution of the unknown parameters in conjunction with the information provided by the sample data. Throughout this book we shall use only the classical methods to estimate unknown population parameters such as the mean and the variance, by computing statistics from random samples, and applying the theory of sampling distributions covered in Sect. 2.3. Bearing in mind the mentioned *parametric* and *nonparametric* statistical models, the methods of statistical inference, i.e. the *estimation* and *hypothesis testing*, can have, depending on a knowledge about a population, parametric or nonparametric form. We start with parametric *point* and *interval estimation*, and then, for the non-parametric case, we present the *histogram* and *kernel probability density estimation*. We also describe certain *bootstrap method* for estimation the confidence intervals of unknown distribution parameters.

2.4.1 Point Estimation

2.4.1.1 Point Estimator and Its Properties

Let's consider the following *parameter estimation* problem. It is convenient to have a general symbol to represent the parameter of interest. By convention, we will use the Greek symbol:

$$\theta \tag{2.4.1.1.1}$$

(*theta*) to represent the parameter. Suppose that an engineer is analyzing the parameter θ, the *torsional resistance* of metal bars. Since variability in torsional resistance is naturally present between the individual bars because of differences in raw material batches, manufacturing processes, and measurement procedures, the engineer is interested in estimating the *mean* torsional resistance of bars. Thus, in practice, the engineer will use the random sample of n bars for which the torsional resistance $x_1, \ldots, x_n$ has been determined, to compute a number:

$$\hat{\theta} = h(x_1, \ldots, x_n) \tag{2.4.1.1.2}$$

that is in some sense a reasonable value (or guess) of an unknown, true mean torsional resistance θ. This number $\hat{\theta}$ is called a *point estimate* of parameter θ. The function $h(x_1, \ldots, x_n)$, when applied to the new sample $x_1, \ldots, x_n$, would yield a different value for $\hat{\theta}$. We, thus, see that estimate is itself a random variable possessing a probability distribution. Thus, if $(X_1, \ldots, X_n)$ is a random sample of size n, the statistic:

$$\hat{\Theta} = h(X_1, \ldots, X_n) \tag{2.4.1.1.3}$$

is called a *point estimator* of θ (please note a hat "^" notation above a symbol of a parameter). After the sample has been selected, $\hat{\Theta}$ takes on a particular numerical value $\hat{\theta}$ called the *point estimate* of θ.

Example 2.4.1.1.1 Let us consider the population represented by the random variable X. The sample mean introduced in section 2.3:

$$\hat{m} = \overline{X} = h(X_1, \ldots, X_n) = \frac{1}{n} \sum_{i=1}^{n} X_i \qquad (2.4.1.1.4)$$

is the reasonable good point estimator of the population mean:

$$E(X) = m \qquad (2.4.1.1.4a)$$

Similarly, the sample variance:

$$\hat{\sigma}^2 = S^2 = h(X_1, \ldots, X_n) = \frac{1}{n} \sum_{i=1}^{n} (X_i - \overline{X})^2 \qquad (2.4.1.1.5)$$

can be used as the point estimator of the population variance:

$$D^2(X) = \sigma^2 \qquad (2.4.1.1.5a)$$

After the sample from population X has been selected, for example (*3, 4, 5*), the numerical value $\bar{x} = 6$ calculated from this sample data, is the *point estimate* of unknown parameter, the population mean $m = E(X)$. Similarly, the numerical value $s^2 = 4.67$, is the *point estimate* of the second unknown population parameter, the variance $\sigma^2 = D^2(X)$.

Mean Square Error of an Estimator

An estimator is not expected to estimate the population parameter without error. In the above example, we do not expect $\overline{X}$ to estimate m exactly, but we certainly hope that it is not far off. Moreover, we may have several different choices for the point estimator of a parameter. For example, if we wish to estimate the mean of a population, we might consider the sample mean, as in the example above, or the sample median, or perhaps the average of the smallest and largest observations in the sample as point estimators.

In order to decide which point estimator of a particular parameter is the best one to use, we need to examine their statistical properties and develop some quality criteria for comparing estimators. One such criterion is the *mean square error* (MSE):

$$MSE(\hat{\Theta}) = E(\hat{\Theta} - \theta)^2 \qquad (2.4.1.1.6)$$

which is the expected squared difference between a point estimator $\hat{\Theta}$ and true parameter value θ. It can be shown that:

$$MSE(\hat{\Theta}) = D^2(\hat{\Theta}) + bias^2(\hat{\Theta}) \qquad (2.4.1.1.7)$$

where the second term is a squared *bias*:

$$bias(\hat{\Theta}) = E(\hat{\Theta}) - \theta \qquad (2.4.1.1.8)$$

of the estimator $\hat{\Theta}$ which is the difference between the expected value of the estimator $\hat{\Theta}$ and the true value of a parameter θ. Thus, the mean squared error of a point estimator $\hat{\Theta}$ is the sum of its variance and squared bias. The mean square error is an important criterion for comparing two estimators. The *relative efficiency* of two estimators $\hat{\Theta}_1$, $\hat{\Theta}_2$ is defined as:

$$\frac{MSE(\hat{\Theta}_1)}{MSE(\hat{\Theta}_2)} \qquad (2.4.1.1.9)$$

Unbiasedness of an Estimator

If the expected value of a point estimator $\hat{\Theta}$ is equal to the parameter θ

$$E(\hat{\Theta}) = \theta \qquad (2.4.1.1.10)$$

we say that an estimator $\hat{\Theta}$ is an *unbiased estimator* for the parameter θ. When an estimator $\hat{\Theta}$ is unbiased, its bias is zero:

$$bias(\hat{\Theta}) = E(\hat{\Theta}) - \theta = 0 \qquad (2.4.1.1.11)$$

Example 2.4.1.1.2 Continuing the previous example, let's check the *unbiasedness* of the two estimators, the sample mean $\overline{X}$ and the sample variance S^2

$$E(\overline{X}) = E\left(\frac{1}{n}\sum_{i=1}^{n}X_i\right) = \frac{1}{n}E\left(\sum_{i=1}^{n}X_i\right) = \frac{1}{n}\sum_{i=1}^{n}E(X_i)$$
$$= \frac{1}{n}(n \cdot m) = m \qquad (2.4.1.1.12)$$

The one before last element in the derivation process above follows directly from a definition of a random sample ($E(X_i) = E(X)$ $i = 1, \ldots, n$). As the bias of $\overline{X}$ is zero:

$$bias(\overline{X}) = E(\overline{X}) - m = m - m = 0 \qquad (2.4.1.1.13)$$

the sample mean $\overline{X}$ is an unbiased estimator of a population mean $m = E(X)$.

Regarding the second estimator S^2, it can be shown that its expected value is:

$$E(S^2) = \sigma^2 - \frac{\sigma^2}{n} = \frac{n-1}{n}\sigma^2 \qquad (2.4.1.1.14)$$

The bias of estimator S^2 is:

$$bias(S^2) = E(S^2) - \sigma^2 = \sigma^2 - \frac{\sigma^2}{n} - \sigma^2 = -\frac{\sigma^2}{n} \qquad (2.4.1.1.15)$$

which means that S^2 underestimates the parameter σ^2 on average.

It is not difficult to notice, that after multiplying S^2 by $n/(n-1)$ we obtain an unbiased estimator S_*^2 of σ^2:

$$S_*^2 = \frac{n}{n-1} \cdot S^2 = \frac{1}{n-1} \sum_{i=1}^{n} (X_i - \overline{X}^2) \qquad (2.4.1.1.16)$$

Its bias is zero:

$$E(S_*^2) = E\left(\frac{n}{n-1} \cdot S^2\right) = \frac{n}{n-1} \cdot E(S^2)$$

$$= \frac{n}{n-1} \cdot \frac{n-1}{n}\sigma^2 = \sigma^2 \qquad (2.4.1.1.17)$$

$$bias(S_*^2) = E(S_*^2) - \sigma^2 = 0 \qquad (2.4.1.1.18)$$

Minimum Variance of an Estimator

It seems natural that if $\hat{\Theta} = h(X_1, \ldots, X_n)$ is to qualify as a good estimator for θ, not only its mean should be close to a true value θ, but also there should be a good probability that any of its observed values $\hat{\theta}$ will be close to θ. This can be achieved by selecting a statistic in such a way that not only is $\hat{\Theta}$ unbiased, but also its variance is as small as possible. Hence, the second desirable property of an estimator is the minimum variance.

Let $\hat{\Theta}$ be an unbiased estimator for θ. It is an ***unbiased minimum-variance estimator*** (**UMVE**) for θ if, for all other unbiased estimators Θ^* of θ from the same sample:

$$D^2(\hat{\Theta}) \leq D^2(\Theta^*) \qquad (2.4.1.1.19)$$

for all θ.

Given two unbiased estimators for a given parameter, the one with smaller variance is preferred because smaller variance implies that observed values of the estimator tend to be closer to its mean, the true parameter value.

Example 2.4.1.1.3 We have seen in Example 2.4.1.1.2 that $\overline{X}$ obtained from a sample of size n is an unbiased estimator for population mean $m = E(X)$. Does the quality of $\overline{X}$ improve as n increases?

Let's calculate the variance of estimator $\overline{X}$:

$$D^2(\overline{X}) = D^2\left(\frac{1}{n}\sum_{i=1}^{n} X_i\right) = \frac{1}{n^2}D^2\left(\sum_{i=1}^{n} X_i\right)$$

$$= \frac{1}{n^2}\sum_{i=1}^{n} D^2(X_i) = \frac{1}{n^2}\cdot n \cdot D^2(X_i) = \frac{\sigma^2}{n} \qquad (2.4.1.1.20)$$

which decreases as n increases. Thus, based on the minimum variance criterion, the quality of $\overline{X}$ as an estimator for $m = E(X)$ improves as n increases.

We may ask further, if based on a fixed sample size n, the estimator $\overline{X}$ is the best estimator for $m = E(X)$ in terms of unbiasedness and minimum variance. That is, if $\overline{X}$ is UMVE for $m = E(X)$?

In order to answer this question, it is necessary to show that the variance of $\overline{X}$, which is equal to σ^2/n, is the smallest among all the unbiased estimators that can be constructed from the sample. A powerful theorem below shows that it is possible to determine the minimum achievable variance of any unbiased estimator obtained from a given sample.

Theorem 2.4.1.1.1 (the Rao-Cramer inequality)

Let $(X_1, \ldots, X_n)$ denote a sample of size n from a population X with pdf $f(x; \theta)$ where θ is the unknown parameter, and let $\hat{\Theta} = h(X_1, \ldots, X_n)$ be an unbiased estimator for θ. Then, the variance of $\hat{\Theta}$ satisfies the inequality:

$$D^2(\hat{\Theta}) \geq \frac{1}{n \cdot E\left[\frac{\partial \ln f(X,\theta)}{\partial \theta}\right]^2} \qquad (2.4.1.1.21)$$

if the indicated expectation and differentiation exist.

An analogous result with pmf $p(x; \theta)$ replacing $f(x; \theta)$ is obtained when X is discrete.

The above theorem gives a lower bound (*Rao-Cramer lower bound* (**RCLB**)) on the variance of any unbiased estimator and it expresses a fundamental limitation on the accuracy with which a parameter can be estimated. We also note that this lower bound is, in general, a function of θ, the true parameter value.

Given any unbiased estimator $\hat{\Theta}$ for parameter θ, the ratio of its RCLB to its variance is called the *efficiency* of $\hat{\Theta}$. The efficiency of any unbiased estimator is thus always less than or equal to 1.

Example 2.4.1.1.4 Let's suppose that the pdf of a population X is normal:

$$f(x, m) = \frac{1}{\sigma\sqrt{2\pi}} \exp\left[-\frac{(x-m)^2}{2\sigma^2}\right] \qquad (2.4.1.1.22)$$

We will show that in this case the sample mean $\overline{X}$ is the UMVE estimator for $m = E(X)$.

We have:

$$\ln f(x, m) = -\ln(\sigma\sqrt{2\pi}) - \frac{(x-m)^2}{2\sigma^2} \qquad (2.4.1.1.23)$$

and the derivative is:

$$\frac{\partial \ln f(x, m)}{\partial m} = \frac{x-m}{\sigma^2} \qquad (2.4.1.1.24)$$

Thus, RCLB is equal to,

$$\frac{1}{nE\left[\frac{\partial \ln f(x,m)}{\partial m}\right]^2} = \frac{1}{nE\left[\frac{x-m}{\sigma^2}\right]^2} = \frac{1}{n \cdot \frac{\sigma^2}{\sigma^4}} = \frac{\sigma^2}{n} \qquad (2.4.1.1.25)$$

In Example 2.4.1.1.3 we have shown that the variance of $\overline{X}$ is equal to:

$$D^2(\overline{X}) = \frac{\sigma^2}{n} \qquad (2.4.1.1.26)$$

which proves that the sample mean $\overline{X}$ is the UMVE for $m = E(X)$, i.e. is the most efficient estimator in normal populations.

It should be noted that sometimes biased estimators are preferable to unbiased estimators because they have smaller mean square error. That is, we may be able to reduce the variance of the estimator considerably by introducing a relatively small amount of bias. As long as the reduction in variance is greater than the squared bias, an improved estimator from a mean square error viewpoint will result. Linear regression analysis is an area in which biased estimators are occasionally used.

Consistency of an Estimator

An estimator $\hat{\Theta}$ is said to be a consistent estimator for θ if, as sample size n increases, the following holds for all ε.

$$\lim_{n\to\infty} P(|\hat{\Theta} - \theta| < \varepsilon) = 1 \qquad (2.4.1.1.27)$$

The consistency condition states that estimator $\hat{\Theta}$ converges, in the sense above, to the true value θ, as sample size increases. It is, thus, a large-sample concept and is a good quality for an estimator to have.

Standard Error of an Estimator

When the numerical value of point estimate of a parameter is reported, it is usually desirable to give some idea of the precision of estimation. Commonly, the *standard error* of an estimator $\hat{\Theta}$ is used which is defined as the standard deviation of its sampling distribution:

$$D(\hat{\Theta}) = \sqrt{D^2(\hat{\Theta})} \qquad\qquad (2.4.1.1.28)$$

If the standard error involves unknown parameters that can be estimated, substitution of those values produces the *estimated standard error*.

Example 2.4.1.1.5 Point estimation of the population mean of salt bag's weight
Continuing Example 2.1.1, we can use the sample mean $\overline{X}$ to estimate the unknown population mean $m = E(X)$ of salt bag's weights based on the random Sample Nr. 1. The point estimate is:

$$\bar{x} = \frac{1}{n}\sum_{i=1}^{n} x_i = 1000.53 \qquad\qquad (2.4.1.1.29)$$

If we can assume that the variance of the salt bag's weight distribution is known, for example $\sigma^2 = 3^2$ (otherwise, it should be estimated from a sample), then the standard error of estimator $\overline{X}$ is:

$$D(\overline{X}) = \frac{\sigma}{\sqrt{n}} = \frac{3}{\sqrt{25}} = 0.6 \qquad\qquad (2.4.1.1.30)$$

Thus, we can report the estimation with precision as follows:

$$\bar{x} = 1000.53 \pm 0.6 \qquad\qquad (2.4.1.1.31)$$

2.4.1.2 Methods of Point Estimation

Often, the estimators of parameters have been those, that appeal to intuition. The estimator $\overline{X}$ certainly seems reasonable as an estimator of a population mean $m = E(X)$. The virtue of S_*^2 as an estimator of $\sigma^2 = D^2(X)$ is underscored through the discussion of unbiasedness in the previous section. But there are many situations in which it is not obvious what the proper estimator should be. As a result, there is much to be learned by the student in statistics concerning different methods of estimation. In this section, we describe the ***method of maximum likelihood***, one of the most important approaches to estimation in all of the statistical inference. This method was developed in the 1920s by a famous British statistician, Sir R. A. Fisher. We introduce the method of maximum likelihood through the example with a discrete distribution and a single parameter. Denote by $(X_1, \ldots, X_n)$, the random sample taken from a discrete probability distribution represented by $p(x; \theta)$, where θ is a single parameter of the distribution, and by $x_1, \ldots, x_n$ the observed values in a sample. The following joint probability:

$$P(X_1 = x_1, \ldots, X_n = x_n | \theta) \qquad\qquad (2.4.1.2.1)$$

which is the probability of obtaining the particular sample values $x_1, \ldots, x_n$ can be written as

$$L(x_1, \ldots, x_n; \theta) = \prod_{i=1}^{n} p(x_i; \theta) \qquad (2.4.1.2.2)$$

The quantity $L(x_1, \ldots, x_n; \theta)$ is called *the likelihood of the sample*, and also often referred to as a *likelihood function*. Note that the variable of the likelihood function is θ, not the x_i.

In the case when X is continuous, we write the likelihood function as:

$$L(x_1, \ldots, x_n; \theta) = \prod_{i=1}^{n} f(x_i; \theta) \qquad (2.4.1.2.3)$$

where $f(x; \theta)$ is a probability density function.

The *maximum likelihood estimator (MLE)* $\hat{\Phi}$ of the parameter θ is the one that maximizes the likelihood function.

$$L(x_1, \ldots, x_n; \hat{\theta}) = \max_{\theta} \ L(x_1, \ldots, x_n; \theta) \qquad (2.4.1.2.4)$$

The algorithm for determining such an estimator can be presented in the following steps.

1. Find the likelihood function L of a sample for a given population distribution.
2. Applying the necessary condition for the existence of an extreme, find the first derivative of the likelihood function L and solve the equation:

$$\frac{\partial L}{\partial \theta} = 0 \qquad (2.4.1.2.5)$$

to obtain the estimator $\hat{\Theta}$ of parameter θ.

Since the function L is always nonnegative and attains its maximum for the same value of as $\ln(L)$, it is sometimes easier to obtain MLE by solving:

$$\frac{\partial \ln(L)}{\partial \theta} = 0 \qquad (2.4.1.2.6)$$

If we have r unknown parameters, then we obtain a system of r equations, in which the left side of each equation is an expression being a partial derivative with respect to one of r parameters. The solution of such a system of r equations is the vector r of estimators of unknown parameters.

It can be shown that estimators obtained by the maximum likelihood method are asymptotically unbiased, effective and consistent.

Example 2.4.1.2.1 Let us consider a population with Poisson distribution:

$$p(x; \theta) = \frac{\theta^x}{x!} e^{-\theta} \quad x = 0, 1, 2, \ldots \tag{2.4.1.2.7}$$

Suppose that a random sample $x_1, \ldots, x_n$ is taken from this distribution to obtain MLE of unknown parameter θ.

The likelihood function is:

$$L(x_1, \ldots, x_n; \theta) = \prod_{i=1}^{n} \frac{\theta^{x_i} e^{-\theta}}{x_i!} = \frac{\theta^{\sum_{i=1}^{n} x_i} e^{-n\theta}}{\prod_{i=1}^{n} x_i!} \tag{2.4.1.2.8}$$

Taking the logarithm of L, we have:

$$\ln L = \sum_{i=1}^{n} x_i \ln \theta - n\theta - \sum_{i=1}^{n} \ln x_i! \tag{2.4.1.2.9}$$

and the derivative of $\ln L$ with respect to θ:

$$\frac{d \ln L}{d\theta} = \frac{\sum_{i=1}^{n} x_i}{\theta} - n \tag{2.4.1.2.10}$$

Setting the derivative to zero, and solving for the parameter θ, we obtain the estimator:

$$\hat{\theta} = \frac{\sum_{i=1}^{n} x_i}{n} \tag{2.4.1.2.11}$$

In the above formula, the x_i are the realizations of the random sample, thus, to obtain an estimator $\hat{\Theta}$, i.e. a statistic, being a random variable, we replace the small x_i in the obtained solution, with the big X_i:

$$\hat{\Theta} = \frac{\sum_{i=1}^{n} X_i}{n} \tag{2.4.1.2.12}$$

Example 2.4.1.2.2 We will obtain MLE of an unknown parameter m in the normal distribution

$$f(x) = \frac{1}{\sqrt{2\pi}\sigma} \exp\left[-\frac{(x-m)^2}{2\sigma^2}\right] \tag{2.4.1.2.13}$$

We assume that σ is known. The likelihood function is:

$$L(x_1, \ldots, x_n; m) = \prod_{i=1}^{n} (2\pi\sigma^2)^{-\frac{1}{2}} \exp\left[-\frac{(x_i-m)^2}{2\sigma^2}\right] =$$

$$= (2\pi\sigma^2)^{\frac{-n}{2}} \exp\left[-\frac{1}{2\sigma^2} \sum_{i=1}^{n} (x_i - m)^2\right] \qquad (2.4.1.2.14)$$

The logarithm of the above expression:

$$\ln L = -\frac{1}{2}n \ln(2\pi\sigma^2) - \frac{1}{2\sigma^2} \sum_{i=1}^{n} (x_i - m)^2 \qquad (2.4.1.2.15)$$

The derivative of $\ln L$ with respect to m is:

$$\frac{d \ln L}{dm} = \frac{1}{\sigma^2} \sum_{i=1}^{n} (x_i - m) \qquad (2.4.1.2.16)$$

Setting the derivative to zero and solving for the parameter m we obtain:

$$\sum_{i=1}^{n} x_i - nm = 0 \qquad (2.4.1.2.17)$$

$$\hat{m} = \frac{\sum_{i=1}^{n} X_i}{n} \qquad (2.4.1.2.18)$$

The MLE estimator of parameter m in the normal distribution, is therefore, the sample mean.

2.4.2 Interval Estimation

Even the most efficient unbiased estimator is unlikely to estimate the population parameter exactly. There are many situations in which it is preferable to determine an interval within which we would expect to find the value of the parameter. Such an interval is called an *interval estimate*. The interval estimate indicates, by its length, the accuracy of the point estimate. Moreover, the interval estimation provides, on the basis of a sample from a population, not only information on the parameter values to be estimated, but also an indication of the level of confidence that can be placed on possible numerical values of the parameters.

Suppose that a sample $(X_1, \ldots, X_n)$ is drawn from a population having probability density function $f(x; \theta)$, θ being the parameter to be estimated. Further suppose that $a_{low}(X_1, \ldots, X_n)$ and $a_{up}(X_1, \ldots, X_n)$ are the two statistics such that $a_{low} < a_{up}$ with probability 1.

The interval:

$$\left(a_{low}, a_{up}\right) \qquad (2.4.2.1)$$

is called a $[100(1 - \alpha)]\%$ *confidence interval* for θ $(0 < \alpha < 1)$ if a_{low} and a_{up} can be selected such that:

$$P(a_{low} < \theta < a_{up}) = 1 - \alpha \qquad (2.4.2.2)$$

The limits a_{low} and a_{up} are called the *lower* and *upper confidence limits* for θ, respectively, and $(1 - \alpha)$ is the *confidence coefficient*. The value of confidence level $(1 - \alpha)$ is generally taken as $0.90, 0.95, 0.99$, and 0.999. Thus, when $\alpha = 0.05$, we have a 95% confidence interval. The wider the confidence interval is, the more confident we can be that the given interval contains the unknown parameter. Of course, it is better to be 95% confident that the average salt bag's weight is between 9.9 and 1.1 kg than to be 99% confident that it is between 9.7 and 1.3 kg! Ideally, we prefer a short interval with a high degree of confidence. Sometimes, restrictions on the size of our sample prevent us from achieving short intervals without sacrificing some of our degree of confidence.

Point and interval estimation represent different approaches to gain information regarding a parameter, but they are related in the sense, that confidence interval estimators are based on point estimators. We know that the estimator $\overline{X}$ is a very reasonable point estimator of $E(X)$. As a result, the important confidence interval estimator of $E(X)$ depends on knowledge of the sampling distribution of $\overline{X}$. We only illustrate the procedure for construction of a confidence interval for a population mean $E(X)$ in the case of normal population with known variance σ^2. This confidence interval is also approximately valid (because of the central limit theorem) regardless of whether or not the underlying population is normal, so long as n is reasonably large $(n > 30)$. However, when the sample is small and σ^2 is unknown, the procedure for constructing a valid confidence interval is based on t distribution (it is not covered in this book—the interested reader is referred to the suggested literature at the end).

Many populations encountered in practice are well approximated by the normal distribution, so this assumption will lead to confidence interval procedures of wide applicability. When the normality assumption is unreasonable, an alternate is to use the *nonparametric procedures*.

Confidence Intervals for Population Mean $m = E(X)$ in $N(x; m, \sigma)$

Let us now consider the interval estimate of $m = E(X)$. If our sample is selected from a normal population **with known** σ, we can establish a confidence interval for m by considering the sampling distribution of $\overline{X}$ and using Theorem 2.3.1 which states that the statistics:

$$U = \frac{\overline{X} - m}{\sigma} \sqrt{n} \qquad (2.4.2.3)$$

has distribution $N(x; 0, 1)$. Suppose we specify that the probability of U being in interval (u_1, u_2) is equal to $1 - \alpha$:

$$P(u_1 < U < u_2) = \Phi(u_2) - \Phi(u_1) = 1 - \alpha \qquad (2.4.2.4)$$

Φ is the cumulative distribution function of the standard normal distribution $N(x; 0, 1)$. From the tables of standard normal distribution, we can find two quantiles u_1, u_2 of $N(x; 0, 1)$ distribution:

$$u_1 = u(\alpha_1), u_2 = u(1 - \alpha_2) \qquad (2.4.2.5)$$

of order α_1 and $1 - \alpha_2$, respectively, where the numbers α_1, α_2 are such that:

$$\alpha_1 + \alpha_2 = \alpha, \ \alpha_1 > 0, \alpha_2 < \alpha \qquad (2.4.2.6)$$

Regarding to the above, we have (see Fig. 2.5):

$$P\left[u(\alpha_1) < \frac{\overline{X} - m}{\sigma}\sqrt{n} < u(1 - \alpha_2)\right] = \Phi(u(1 - \alpha_2)) - \Phi(u(\alpha_1))$$

$$= 1 - \alpha_2 - \alpha_1 = 1 - (\alpha_2 + \alpha_1) = 1 - \alpha \qquad (2.4.2.7)$$

Solving the above with respect to m, we get :

$$P\left(\overline{X} - u(1 - \alpha_2)\frac{\sigma}{\sqrt{n}} < m < \overline{X} - u(\alpha_1)\frac{\sigma}{\sqrt{n}}\right) = 1 - \alpha \qquad (2.4.2.8)$$

which leads to the interval:

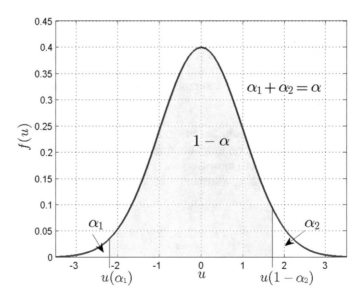

Fig. 2.5 Pdf of U statistics with the two quantiles marked

$$\left(\overline{X} - u\left(1 - \frac{\alpha}{2}\right) \cdot \frac{\sigma}{\sqrt{n}}, \overline{X} - u\left(\frac{\alpha}{2}\right) \cdot \frac{\sigma}{\sqrt{n}}\right) \qquad (2.4.2.9)$$

Depending on the way of choosing α_1, α_2, we can create infinitely many confidence intervals. It can be shown that the shortest confidence interval is obtained if we assume:

$$\alpha_1 = \alpha_2 = \frac{\alpha}{2} \qquad (2.4.2.10)$$

Based on the symmetry of the $N(x; 1, 0)$ distribution:

$$-u\left(\frac{\alpha}{2}\right) = u\left(1 - \frac{\alpha}{2}\right) \qquad (2.4.2.11)$$

we obtain the following confidence interval:

$$\left(\overline{X} - u\left(1 - \frac{\alpha}{2}\right) \cdot \frac{\sigma}{\sqrt{n}}, \overline{X} + u\left(1 - \frac{\alpha}{2}\right) \cdot \frac{\sigma}{\sqrt{n}}\right) \qquad (2.4.2.12)$$

The obtained interval is a *random interval* and must be interpreted carefully. The mean m, although unknown, is nevertheless deterministic and it either lies in an interval or it does not. The probability that this random interval covers the distribution's true mean m is $1 - \alpha$. Based upon the given sample values, we get the *observed interval*.

The obtained interval is symmetrical with respect to $\bar{x}$, and has the length:

$$l_n = 2u\left(1 - \frac{\alpha}{2}\right) \cdot \frac{\sigma}{\sqrt{n}} \qquad (2.4.2.13)$$

For a fixed sample size and confidence coefficient, the length is constant. The greater the sample size, the shorter the confidence interval, i.e. the greater the accuracy.

The $[100(1 - \alpha)]\%$ confidence interval provides an estimate of the accuracy of our point estimate $\bar{x}$. If $m = E(X)$ is actually the center value of the interval, then $\bar{x}$ estimates m without error. The size of this error will be the absolute value of the difference between m and $\bar{x}$, and we can be $[100(1 - \alpha)]\%$ confident that this difference will not exceed $l_n/2 = u(1 - \alpha/2) \cdot \sigma/\sqrt{n}$, being the maximum error of estimation.

The accuracy of the estimation can be quantitatively described by the so-called *relative precision*:

$$relprec = \frac{l_n/2}{\bar{x}} \cdot 100\% \qquad (2.4.2.14)$$

In practice, often it is assumed that statistical inference based on a confidence interval is "safe", when the relative precision does not exceed 5% and if it exceeds 10%, we should then increase the sample size or assume a lower confidence coefficient.

Determination of the Minimum Sample Size
Frequently, we wish to know how large a sample is necessary to ensure that the error in estimating m will be less than a specified amount $l_n = 2d$. In our case of normal population with known σ, using the obtained above formula for interval length we have:

$$2u\left(1 - \frac{\alpha}{2}\right)\frac{\sigma}{\sqrt{n}} \leq 2d \qquad (2.4.2.15)$$

from which we calculate the minimum of sample size:

$$n \geq \frac{u^2\left(1 - \frac{\alpha}{2}\right)\sigma^2}{d^2} \qquad (2.4.2.16)$$

Example 2.4.2.1 Interval estimation of the average weight of a salt bag
Let's return to Example 2.1.1, regarding the evaluation of the average weight of salt bags in a certain factory. Assume that the weight X of the salt bag has a normal distribution with a known standard deviation $\sigma = 3$. At the assumed confidence level $1 - \alpha = 1 - 0.05 = 0.95$, we determine the confidence interval for the unknown mean weight:

$$\left(\overline{X} - u\left(1 - \frac{\alpha}{2}\right)\frac{\sigma}{\sqrt{n}}, \overline{X} + u\left(1 - \frac{\alpha}{2}\right)\frac{\sigma}{\sqrt{n}}\right)$$
$$= \left(1000.53 - 1.96\frac{3}{\sqrt{25}}, 1000.53 + 1.96\frac{3}{\sqrt{25}}\right)$$
$$= (999.35, 1001.70) \qquad (2.4.2.17)$$

where the value of the quantile of this distribution read from tables of the standard normal distribution, for the assumed confidence level equals to $u(1 - 0.05/2) = 1.96$. So, we have 95% certainty that the interval $(999.35, 1001.70)$ covers the unknown mean salt bag weight. The length of this interval is:

$$2u\left(1 - \frac{\alpha}{2}\right) \cdot \frac{\sigma}{\sqrt{n}} = 2 \cdot 1.96 \cdot \frac{3}{\sqrt{25}} = 2.35 \qquad (2.4.2.18)$$

The relative precision of the above estimation is:

$$relprec = \frac{1.175}{1000.53} \cdot 100\% = 0.117\% \qquad (2.4.2.19)$$

Now, let's assume that we want even greater accuracy, i.e. a shorter interval with length:

$$2d = 1.5 < 2.35 \tag{2.4.2.20}$$

The minimum sample size will be:

$$n \geq \frac{u_{1-\alpha}^2 \sigma^2}{d^2} = \frac{1.96^2 \cdot 3^2}{(0.75)^2} \approx 61.47 \tag{2.4.2.21}$$

It would be necessary to increase the sample size by at least 37 ($37 = 62 - 25$) elements to achieve the desired, more accurate estimation of the mean salt bag's weight.

2.4.3 Bootstrap Estimation

The *bootstrap* was first published by Bradley Efron in "*Bootstrap methods: another look at the jackknife*" (The Annals of Statistics, 7(1), pp. 1–26 (1979)). In statistics, *bootstrapping* is any method that relies on random sampling with replacement. The idea behind bootstrap is to use the data of a sample study at hand as a "*surrogate population*" for the purpose of approximating the sampling distribution of a statistic.

Let us develop now some mathematical notations for explaining the **bootstrap approach**. If we have a random sample data of size n:

$$(x_1, \ldots, x_n) \tag{2.4.3.1}$$

resampling with replacement from the sample data at hand creates a large number of "phantom samples":

$$(x_1^*, \ldots, x_n^*) \tag{2.4.3.2}$$

known as *bootstrap samples* (we denote a resample of size n by adding a star to the symbols).

Suppose now the population mean is the target of our study (in general the population parameter θ). The corresponding sample statistic computed from the sample data is:

$$\hat{\theta} = h(x_1, \ldots, x_n) = \bar{x} = \frac{1}{n} \sum_{i=1}^{n} x_i \tag{2.4.3.3}$$

Similarly, just as $\bar{x}$ is the sample mean of the original data, we write $\bar{x}^*$ (in general, $\hat{\theta}^*$) for the mean of the resampled data. We then compute:

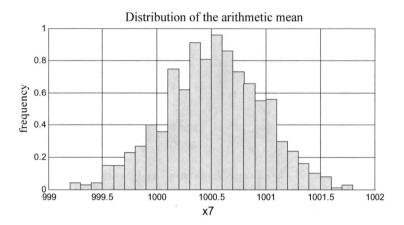

Fig. 2.6 Bootstrap distribution of the sample mean for $N = 1000$

$$\hat{\theta}_1^*, \hat{\theta}_2^*, \ldots, \hat{\theta}_N^* \tag{2.4.3.4}$$

using the same computing formula as the one used for $\hat{\theta}$, but now base it on N (usually a few thousand) different bootstrap samples (each of size n). From the values $\hat{\theta}_1^*, \hat{\theta}_2^*, \ldots, \hat{\theta}_N^*$ of the statistic computed on the subsequent bootstrap samples, the histogram is created (referred here to as the *bootstrap distribution* of the statistic) (see Fig. 2.6).

The primary application of bootstrap is approximating the standard error of a sample estimate. One defines the *bootstrap standard error* as:

$$SE_B(\hat{\theta}) = \sqrt{\frac{1}{N} \sum_{i=1}^{N} (\hat{\theta}_i^* - \hat{\theta})^2} \tag{2.4.3.5}$$

The second applications of bootstrap are the bootstrap confidence intervals. We describe here one of such methods, the *bootstrap percentile method*.

Suppose one settles for 1000 bootstrap replications of $\hat{\theta}$, denoted by

$$(\hat{\theta}_1^*, \hat{\theta}_2^*, \ldots, \hat{\theta}_{1000}^*) \tag{2.4.3.6}$$

After ranking from bottom to top, let us denote these bootstrap values as:

$$(\hat{\theta}_{(1)}^*, \hat{\theta}_{(2)}^*, \ldots, \hat{\theta}_{(1000)}^*) \tag{2.4.3.7}$$

Then, the *bootstrap percentile confidence interval* at 95% level of confidence is:

$$(\hat{\theta}_{(25)}^*, \hat{\theta}_{(975)}^*) \tag{2.4.3.8}$$

(i.e. the 25th and 97th percentiles). It should be pointed out that this method requires the symmetry of the sampling distribution of θ around $\hat{\theta}$.

Example 2.4.3.1 Let us continue the example of estimating the mean salt bag's weight (Example 2.1.1). We determine the confidence intervals at 95% level of confidence for the mean salt bag's weight using the percentile method. From the Sample No. 1 we generate $N = 1000$ bootstrap samples, and for each of them we calculate the sample mean $\bar{x}^*$. We then sort the 1000 values in ascending order and find the 25th and 97th percentiles. Figure 2.6 shows the histogram of the bootstrap sample means $\bar{x}^*$, i.e. the bootstrap distribution of sample mean. The bootstrap confidence interval for the mean salt bag's weight at 95% level of confidence is:

$$(999.59, \quad 1001.39) \tag{2.4.3.9}$$

2.4.4 Nonparametric Density Estimation

In this section, instead of assuming a parametric model for the distribution (e.g. normal distribution with unknown expectation and variance), we only assume that the probability density function exists and is suitably smooth (e.g. differentiable). It is then possible to estimate the unknown probability density function $f(x)$. We assume that we have a random sample $(X_1, \ldots, X_n)$ where $X_i \sim F$, and F denotes an unknown cumulative distribution function.

The goal is to estimate the distribution F. In particular, we are interested in estimating the density $f = dF/dx$, assuming that it exists. The resulting estimator will be denoted as $\hat{f}(x)$.

In this section, we consider two such approaches: the *histogram* and *kernel estimators*.

2.4.4.1 Histogram Estimation Method

The *histogram* is the oldest density estimator. We need to specify an "*origin*" x_0 and the class width h (also called a *bandwidth*) for the specifications of the intervals:

$$c_k = [x_0 + kh, x_0 + (k+1)h) \quad k = 0, \pm 1, \pm 2, \ldots. \tag{2.4.4.1.1}$$

for which the histogram counts the number of observations falling into each c_k. We then plot the histogram such that the area of each bar is proportional to the number of observations falling into the corresponding class (interval c_k).

In searching for the optimal density estimator, we can minimize *the mean squared error $MSE(x)$* at a point x,

$$MSE(x) = \left(E\left[\hat{f}(x) - f(x)\right]\right)^2 + Var(f(x)) \qquad (2.4.4.1.2)$$

but, then we are confronted with a bias-variance trade-off. Instead of optimizing the mean squared error at a point x, the *integrated Mean Square Error* (MISE) defined by the following formula is used:

$$MISE = E[\int_{-\infty}^{+\infty} (\hat{f}(x) - f(x))^2 dx] \qquad (2.4.4.1.3)$$

Minimizing the MISE, results in the "global" asymptotically optimal bandwidth h which depends on f'' which is unknown. However, if we assume that f is normal, i.e. $N(x; m, \sigma)$, we can obtain a formula for optimal bandwidth:

$$h = 3.486 \cdot \sigma \cdot n^{-1/2} \qquad (2.4.4.1.4)$$

where the unknown value of the σ is estimated from the sample by the sample standard deviation s. This gives quite good results also for the distributions differing significantly from the normal distribution.

Example 2.4.4.1.1 Let's return to Example 2.1.1 and assume that we do not know the functional form of the salt bag's weight probability density function.

We will construct the histogram estimator based on the Sample Nr. 1. The bandwidth h is calculated as (a sample standard deviation for sample Nr 1 is appr. 2.27):

$$h = 3.486 \cdot \sigma \cdot \sqrt{n} = 3.486 \cdot 2.27/\sqrt{25} = 1.579 \qquad (2.4.4.1.5)$$

The calculations of histogram estimator are shown in Table 2.4 for the selected range (990.00, 1012.109) which was divided into *14* intervals with the length $h = 1.579$. Figure 2.7 presents the plot of the constructed histogram estimator.

2.4.4.2 Kernel Estimation Method

As in the histogram, the relative frequency of observations falling into a small region can be computed. The density function $f(x)$ at a point x can be represented as:

$$f(x) = \lim_{h \to 0} \frac{1}{2h} P(x - h < X \le x + h) \qquad (2.4.4.2.1)$$

The simplest estimator can be constructed (neglecting the limit) by replacing probabilities with relative frequencies (|.| means the number):

Table 2.4 Calculations for histogram estimator

	From	To	Numerical amount	$\hat{f}_{25}(x)$
1	990.000	991.579	0	0.00
2	991.579	993.158	0	0.00
3	993.158	994.738	0	0.00
4	994.738	996.317	2	0.05
5	996.317	997.896	1	0.03
6	997.896	999.475	4	0.10
7	999.475	1001.055	6	0.15
8	1001.055	1002.634	8	0.20
9	1002.634	1004.213	2	0.05
10	1004.213	1005.792	2	0.05
11	1005.792	1007.372	0	0.00
12	1007.372	1008.951	0	0.00
13	1008.951	1010.530	0	0.00
14	1010.530	1012.109	0	0.00

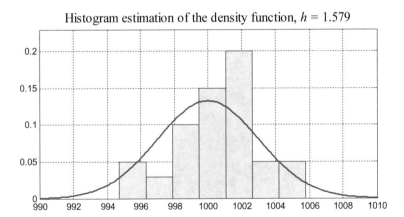

Fig. 2.7 Histogram estimator for salt bag's weight distribution based on Sample Nr. 1. The plot of $N(x; 1000, 3)$ superimposed

$$\hat{f}(x) = \frac{1}{2h} |\{i : X_i \in (x - h, \ x + h)\}| \qquad (2.4.4.2.2)$$

We can represent this estimator in an alternative way:

$$\hat{f}(x) = \frac{1}{nh} \sum_{i=1}^{n} K\left(\frac{x - X_i}{h}\right) \qquad (2.4.4.2.3)$$

where

$$K(x) = \begin{cases} 1/2 & \text{if } |x| \le 1 \\ 0 & \text{otherwise} \end{cases} \qquad (2.4.4.2.4)$$

is a *rectangle weight function*. Similar to histogram, this estimator is only piecewise constant.

Instead of the rectangle weight function we can choose a general, typically smoother function $K(.)$ (called kernel function), satisfying the following conditions:

$$K : R \to [0, +\infty) \qquad (2.4.4.2.5)$$

$$\int\limits_{-\infty}^{+\infty} K(x)dx = 1 \quad \underset{x \in R}{\forall} \; K(-x) = K(x) \qquad (2.4.4.2.6)$$

In this way we have the definition of the *kernel density estimator*:

$$\hat{f}(x) = \frac{1}{nh} \sum_{i=1}^{n} K\left(\frac{x - X_i}{h}\right) \qquad (2.4.4.2.7)$$

Asymptotically optimal (i.e. minimizing the mean square error) kernel is the *Epanechnikov kernel*:

$$K_e(u) = \begin{cases} \frac{3}{4}(1 - u^2) & x \in B = [-1, 1] \\ 0 & u \in R \backslash B \end{cases} \qquad (2.4.4.2.8)$$

where

$$u = \frac{x - X_i}{h} \qquad (2.4.4.2.9)$$

The optimal bandwidth h depends on the form of the kernel and the unknown density function f. Assuming that f is the density of the normal distribution $N(x; m, \sigma)$, the following formula for bandwidth is obtained:

$$h_e = 1.05\sigma \, n^{-1/5} \qquad (2.4.4.2.10)$$

The σ parameter can be estimated from the sample standard deviation s. It turns out that such an estimation of the parameter h gives quite good results also for distributions differing significantly from the normal distribution.

Study of the asymptotic mean square error shows that the use of another kernel, the *Gaussian kernel*:

$$K_g(u) = \frac{1}{\sqrt{2\pi}} \exp\left(-\frac{u^2}{2}\right) \qquad (2.4.4.2.11)$$

leads to an increase in the mean square error by only a few percent and the optimal bandtwidth is practically the same as at Epanechnikov.

Example 2.4.4.2.1 Continuing the previous example, we will construct now a kernel estimator of the unknown density function using the Gaussian kernel. The bandwidth h is calculated according to Formula (2.4.4.2.10):

$$h_e = 1.05 \cdot 2.27 \cdot (25)^{-1/5} = 1.249 \qquad (2.4.4.2.12)$$

Let's calculate for example the value of the kernel estimator at point $x = 1000$:

$$\hat{f}(1000) = \frac{1}{25 \cdot 1.249\sqrt{2\pi}} \left[\exp\left(-\frac{(1000 - 1000.33)^2}{2 \cdot (1.249)^2}\right) \right.$$
$$+ \cdots + \exp\left(-\frac{(1000 - 1001.72)^2}{2 \cdot (1.249)^2}\right) \left.\right]$$
$$= 0.0128 \cdot \left[\exp\left(-\frac{0.111}{3.12}\right) + \cdots + \exp\left(-\frac{2.970}{3.12}\right) \right]$$
$$= 0.0128 \cdot \left[\exp(-0.036) + \cdots + \exp(-0.952) \right]$$
$$= 0.0128 \cdot [0.965 + \cdots + 0.386] = 0.0128 \cdot 11.010 = 0.1409$$
$$(2.4.4.2.13)$$

In a similar way, we calculate the values of the kernel estimator at the remaining points from the range (*990.00, 1012.109*). We performed calculations of the values at 350 points in the above range with a step of 0.6. Figure 2.8 shows a plot of the kernel estimator obtained for the value of $h = 1.249$.

2.5 Tests of Hypotheses

Besides the estimation of parameters presented in the previous section, often the problem of inference about a population relies on the formation of a data-based *decision procedure*. For example, a medical researcher may decide on the basis of experimental evidence whether smoking cigarettes increases the risk of cancer in humans; an engineer might have to decide on the basis of sample data whether there is a difference between the accuracy of two kinds of gauges produced. Similarly, a psychologist might wish to collect appropriate data to enable him to decide whether a person's personality type and intelligence quotient are independent variables.

In each of the above cases, the person postulates or conjectures something about an investigated situation. In addition, each must involve the use of experimental data and decision-making that is based on the data. The conjecture can be put in the form

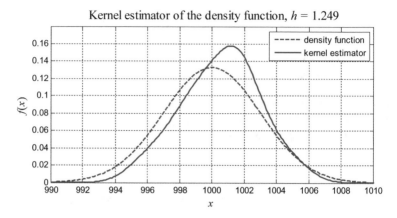

Fig. 2.8 Kernel estimator for salt bag's weight distribution based on Sample Nr. 1. The plot of $N(x;$ 1000, 3) superimposed (dashed line)

of a *statistical hypothesis*. Procedures that lead to the acceptance or rejection of such statistical hypotheses are called *hypothesis testing* and they comprise a major area of statistical inference. Hypothesis testing utilizes a properly constructed statistic known as the *test statistic* $T_n = h(X_1, \ldots, X_n)$.

Formally, a *statistical hypothesis* is an assertion or conjecture concerning the one or more populations. Since we use probability distributions to represent populations, a statistical hypothesis may also be thought of as a statement about the probability distribution of a random variable X. The hypothesis will usually involve one or more parameters of this distribution. For example, assume that population salt bag's weight from our Example 2.1.1 is a random variable X with normal distribution $N(x; m, \sigma)$ and known variance σ^2. The hypothesis:

$$H_0: \ m = 1000 \tag{2.5.1}$$

is the example of a *parametric hypothesis*. If a functional form of a distribution of a random variable X in unknown, the hypothesis

$$H_0: \text{ distribution of } X \text{ is normal} \tag{2.5.2}$$

is the example of a *nonparametric hypothesis*.

The truth or falsity of a statistical hypothesis is never known with absolute certainty unless we examine the entire population which is impractical in most situations. Instead, we take a random sample from the population of interest and use the sample data to provide evidence that either supports or does not support the hypothesis. In the last case, such a postulated hypothesis is rejected. It should be made clear that such a decision must be done with the awareness of the probability of a wrong conclusion which will be described later.

2.5.1 General Concepts

To illustrate the general concepts of hypothesis testing, consider the problem of assessing the mean weight of salt bags produced in a certain factory (Example 2.1.1, Sample 1). Suppose it can be assumed that salt bag's weight X has a normal distribution $N(x; m, \sigma)$, where $\sigma = 15$. Specifically, we are interested in deciding whether the mean salt bag's weight $m = E(X)$ is 1000 g or it is greater. We may express this formally as

$$H_0 : \ m = 1000 \tag{2.5.1.1}$$

$$H_1 : \ m > 1000 \tag{2.5.1.2}$$

The statement $H_0 : \ m = 1000$ is called the *null hypothesis*, and the statement $H_1 :$ $m > 1000$—the *alternative hypothesis*. It is the example of *one-sided* alternative hypothesis. However, in some situations, we may wish to formulate the *two-sided* alternative hypothesis, $H_1 : \ m \neq 1000$.

Testing the hypothesis involves taking a random sample, computing a *test statistic* from a sample data, and then, using a test statistic to make a decision about a null hypothesis. Referring to Example 2.1.1, a value of the sample mean $\bar{x}$ that falls close to the hypothesized value of $m = 1000$ g is the evidence that the true mean m is really 1000 g, that is, such evidence supports the null hypothesis H_0. On the other hand, a sample mean that is considerably different from 1000 g is the evidence in support of the alternative hypothesis H_1. Thus, the sample mean is the test statistic in this case (i.e. $T_n = \overline{X}$). The sample mean can take on many different values. Suppose, that if $\bar{x} > 1003$ we will reject the null hypothesis in favor of the alternative: the values of $\bar{x}$ that are greater than 1003 constitute the *critical region* for the test:

$$K_0 = \{x : \ \overline{X} > c = 1003\} \tag{2.5.1.3}$$

All the values that are less than 1003

$$K_1 = \{x : \ \overline{X} < 1003\} \tag{2.5.1.4}$$

form the *acceptance region* for which we will fail to reject the null hypothesis. It should be noted that the critical and acceptance regions correspond with the form of alternative hypothesis stated above.

The boundaries between the critical and the acceptance regions are called the *critical values* (here, $c = 1003$). We reject H_0 in favor of H_1 if the test statistic falls in the critical region and fail to reject H_0 otherwise.

This decision procedure can lead to either of two wrong conclusions. For example, the true mean salt bag weight could be equal to 1000 g. However, for the randomly selected sample that are tested, we could observe a value of the test statistic $\bar{x}$ that falls into the critical region. We would then reject the null hypothesis H_0 in favor of the alternative H_1 when, in fact, H_0 is really true. This type of wrong conclusion is called a *type I error*.

Now, suppose that the true mean salt bag weight is different from 1000 g, yet the sample mean falls in the acceptance region. In this case, we would fail to reject H_0 when it is false. This type of wrong conclusion is called a *type II error*.

Summarizing, rejecting the null hypothesis H_0 when it is true is defined as a **type I error**. Failing to reject the null hypothesis when it is false is defined as *a **type II error***.

Because our decision is based on random variables, probabilities can be associated with the type I and type II errors.

The probability of type I error is denoted by α and defined as

$$\alpha = P(T_n \in K_0|H_0) \tag{2.5.1.5}$$

The type I error probability is called the *significance level of the test*.

The probability of type II error, which we will denote by β is defined as

$$\beta = P(T_n \in K_1|H_1) \tag{2.5.1.6}$$

To calculate β, we must have a specific alternative hypothesis; that is, we must have a particular value of m. For example, suppose that it is important to reject the null hypothesis $H_0 : m = 1000\,\text{g}$ whenever the (true) mean salt bag's weight is greater than $m = 1009$ g. We could calculate the probability of a type II error β for the values $m = 1009\,\text{g}$ and use this result to tell us something about how the test procedure would perform. Specifically, how will the test procedure work if we wish to detect, that is, reject H_0, for a salt bag's weight mean value of $m = 1009\,\text{g}$?

Let's calculate the two probabilities of error. The probability of the type I error is equal to the area that have been shaded in the right tail of the normal distribution $N(1000.15/\sqrt{25})$ in Fig. 2.9. We may find this probability as

$$\alpha = P(\overline{X} > 1003|H_0) = P(Z > 1|H_0) = \int_{1}^{+\infty} \phi(x)$$

$$= 1 - \Phi(1) = 0.159 \tag{2.5.1.7}$$

because the standardized value that corresponds to the critical value 1003 is $z = (1003 - 1000)/(15/\sqrt{25}) = 1$, ϕ and Φ symbols denote the probability density and cumulative distribution function of the standard normal distribution, respectively.

Now, when the alternative hypothesis H_1 is true (i.e. $m = 1009\,\text{g}$), a type II error will be committed if the value of test statistic $\bar{x}$ falls into the acceptance region, i.e. $\bar{x} < 1003$. The probability of the type II error is equal to the area that have been shaded in the left tail of the normal distribution $N(1009.15/\sqrt{25})$ in Fig. 2.9. Therefore, referring to Fig. 2.9, we find that

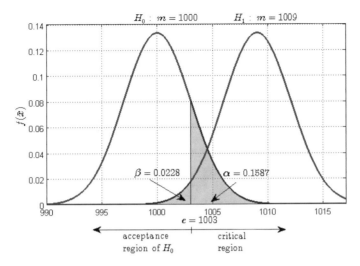

Fig. 2.9 Illustration of the type I and II probability of error ($c = 1003$)

$$\beta = P(\overline{X} < 1003|H_1) = P(Z < -2|H_1) = \int_{-\infty}^{-2} \phi(x)$$

$$= \Phi(-2) = 0.023 \qquad (2.5.1.8)$$

because the standardized value that corresponds to the critical value 1003 is now $z = (1003 - 1009)/(15/\sqrt{25}) = -2$.

The obtained results imply that about 15% of all random samples would lead to rejection of the null hypothesis $H_0 :\ m = 1000\,\mathrm{g}$ when the true mean weight is really 1000 g. If the true value of the mean weight is $m = 1009\,\mathrm{g}$, the probability that we will fail to reject the false null hypothesis is 0.023 (2.3%).

From inspection of Fig. 2.10, we notice that we can reduce the probability of type I error α by widening the acceptance region. For example, if we make the critical value $c = 1005$, the value of α decreases to

$$\alpha = P(\overline{X} > 1005|H_0) = P(Z > 1.666|H_0) = 0.047 \qquad (2.5.1.9)$$

$\left(z = (1005 - 1000)/(15/\sqrt{25}) = 1.666\right)$. Unfortunately, at the same time the value of probability of β increases to

$$\beta = P(\overline{X} < 1005|H_1) = P(Z < -1.333|H_1) = 0.091 \qquad (2.5.1.10)$$

where the standardized value that corresponds to the critical value 1005 is now $z = (1005 - 1009)/(15/\sqrt{25}) = -1.333$.

The following conclusions can be formulated from the obtained above results.

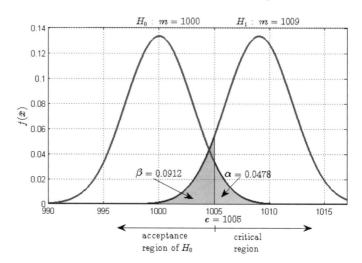

Fig. 2.10 Illustration of the type I and II probability of error ($c = 1005$)

- The probability of a type I error α can always be reduced by appropriate selection of the critical value(s) (a critical region).
- Type I and type II errors α and β are related. A decrease in the probability of one type of error always results in an increase in the probability of the other, provided that the sample size n does not change.
- Moreover, it can be shown that an increase in sample size n will generally reduce both α and β, provided that the critical values are held constant.

Generally, a statistician controls the type I error probability α (i.e. the significance level α) when he selects the critical values. Because he can directly control the probability of wrongly rejecting H_0, this rejection of the null hypothesis is a *strong* conclusion.

On the other hand, controlling the probability of type II error β is not so easy: probability β is a function of both the sample size and the extent to which the null hypothesis H_0 is false. Therefore, in the so-called **significance tests**, rather than saying we "*accept H_0*", we prefer the terminology "*fail to reject H_0*". Failing to reject H_0 implies that we have not found sufficient evidence to reject H_0 (i.e. it does not necessarily mean that there is a high probability that H_0 is true—it may simply mean that more data are required to reach a strong conclusion).

A Power of Statistical Test

An important concept that we will make use of is the power of a statistical test. A *power of a statistical test* based on a critical region K_0 is the probability of rejecting the null hypothesis H_0 when the alternative hypothesis H_1 is true:

$$power(K_0, H_1) = P(T \in K_0|H_1)$$
$$= 1 - P(T \in K_1|H_1) = 1 - \beta \qquad (2.5.1.11)$$

The power can be interpreted as the probability of correctly rejecting a false null hypothesis. We often compare statistical tests by comparing their power properties. In our example of testing $H_0 : m = 1000$ based on the critical region $K_0 = \{x : \overline{X} > 1003\}$, we found that $\beta = 0.023$, so the power of this test is $1 - \beta = 1 - 0.023 = 0.977$.

Power is a measure of the *sensitivity* of a statistical test, where by sensitivity we mean the ability of the test to detect differences. In this case, the sensitivity of the test for detecting the difference between the mean salt bag's weight of 1000 and 1009 g is 0.977. That means if the true mean weight is really 1009 g, this test will correctly reject $H_0 : m = 1000$, and "detect" this difference 97.7% of the time.

A test of any hypothesis such as:

$$H_0 : m = m_0$$
$$H_1 : m < m_0 \text{ or } H_1 : m > m_0 \quad (2.5.1.12)$$

is called *one-tailed* (or *one-sided*) test. If the alternative hypothesis is $H_1 : m < m_0$ ($H_1 : m > m_0$), the critical region should lie in the lower (upper) tail of the distribution of the test statistic.

A test of any hypothesis such as:

$$H_0 : m = m_0$$
$$H_1 : m \neq m_0 \quad (2.5.1.13)$$

is called a *two-tailed* (*two-sided*) test, because it is important to detect differences from the hypothesized value m_0 of the mean that lie on either side of m_0. In such a test, the critical region is split into two parts, with usually equal probability placed in each tail of the distribution of the test statistic.

In constructing hypotheses, we will always state the null hypothesis H_0 as an equality. so that the probability of type I error α can be controlled at a specific value. The alternative hypothesis H_1 might be either one-sided or two-sided, depending on the conclusion to be drawn if H_0 is rejected. If the objective is to make a claim involving statements such as greater than, superior to, less than, worse than and so forth, a one-sided alternative is appropriate. If no direction is implied by the claim, a two-sided alternative should be used.

At last, it should be noted that there is a close relationship between the test of a hypothesis about any parameter, say, m, and the confidence interval for this parameter m. If (m_{low}, m_{up}) is a $100(1 - \alpha)\%$ confidence interval for the parameter m, the test on significance level α of the null hypothesis like $H_0 : m = m_0$, will lead to the rejection of H_0 if and only if a parameter m is not in the $100(1 - \alpha)\%$ confidence interval (m_{low}, m_{up}). In this sense, hypothesis tests and confidence intervals are equivalent procedures.

In this section we have developed the general philosophy for hypothesis testing. We recommend the following sequence of steps in applying hypothesis testing methodology.

General Procedure for Hypothesis Testing in Significance Tests

- State the null hypothesis H_0 about the investigated population depending on the problem.
- Specify the appropriate alternative hypothesis H_1.
- Choose the significance level α (i.e. the probability of type I error).
- Determine the appropriate test statistic.
- Determine the critical (rejection) region for the statistic.
- For the random sample from an investigated population, compute the value of test statistic.
- Decide whether or not H_0 should be rejected and report that in the problem context.

The above stated sequence of steps will be illustrated in subsequent two sections. We would like to notice that we do not provide statistical tables in this book since R environment provides the relevant critical values to construct critical regions.

2.5.2   Parametric Tests

The hypothesis-testing procedures discussed in this section are based on the assumption that we are working with random samples from *normal populations*. Traditionally, we have called these procedures *parametric methods* because they are based on a particular parametric family of distributions—in this case, the normal. Later, in the next section, we describe procedures called *nonparametric* or *distribution-free methods*, which do not make assumptions about the distribution of the underlying population other than that it is continuous. However, the assumptions of normality often cannot be justified and we do not always have quantitative measurements. In such cases, the nonparametric methods are used with increasing frequency by data analysts.

In this section we will assume that a random sample $(x_1, \ldots, x_n)$ has been taken from the **normal population** $N(x; m, \sigma)$. We will consider tests of hypotheses on a single population parameters such as mean and variance and then, we extend those results to the case of two independent populations: we present hypotheses tests for a difference in means and variances.

2.5.2.1   Hypothesis Tests on the Population Mean

Suppose that we wish to test the hypotheses:

$$H_0 : \ m = m_0$$
$$H_1 : \ m \neq m_0 \qquad (2.5.2.1.1)$$

where m_0 is a specified constant. Here, we will consider three cases, depending on the knowledge of the value of the standard deviation σ in the normal population $N(x; m, \sigma)$ under study:

- σ known,
- σ unknown,
- large sample.

Case 1. Standard Deviation σ is Known

If the hypothesis is true, then the distribution of the sample mean $\overline{X}$ is normal distribution with the mean value m_0 and the standard deviation $\sigma/\sqrt{n}$, and we could construct the critical region based on the computed value of the sample mean $\overline{X}$, as in the previous section. It is usually more convenient to standardize the sample mean and use the test statistic:

$$U = \frac{\overline{X} - m_0}{\sigma/\sqrt{n}} \qquad (2.5.2.1.2)$$

that has a normal distribution $N(x; 0, 1)$ which follows from Theorem 2.3.1. Hence, the expression:

$$P\left(-u\left(1 - \frac{\alpha}{2}\right) < \frac{\overline{X} - m_0}{\sigma/\sqrt{n}} < u\left(1 - \frac{\alpha}{2}\right)\right) = 1 - \alpha \qquad (2.5.2.1.3)$$

can be used to write an appropriate nonrejection region which follows from the symmetry of the normal distribution $N(x; 0, 1)$, where $u(1 - \alpha/2)$ is a quantile of order $(1 - \alpha/2)$ of this distribution (see Fig. 2.11).

Clearly, a sample producing a value of the test statistic U that falls in the tails of the distribution of U would be unusual if $H_0 : \ m = m_0$ is true. Therefore, it is an indication that H_0 is false. Thus, we should reject H_0 if the observed value of the test statistic U falls in the *critical (rejection) region*:

$$K_0 = \left(-\infty, -u\left(1 - \frac{\alpha}{2}\right)\right) \cup \left(u\left(1 - \frac{\alpha}{2}\right), +\infty\right) \qquad (2.5.2.1.4)$$

Note that the probability is α that the test statistic U will fall in the critical region. It should be kept in mind that, formally, the critical region is designed to control α, the probability of type I error.

Example 2.5.2.1.1 Hypothesis test for the population mean salt bag's weight (σ is known)

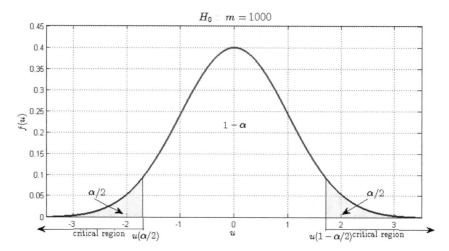

Fig. 2.11 Plot of the distribution of the standardized mean test statistic with the critical region shaded

Let us consider the already investigated problem of the assessment of the mean weight of salt bags produced in a certain factory (Example 2.1.1, Sample 1). Assuming that the weight's distribution is approximately normal with the standard deviation $\sigma = 3$ known, we've already found the $100 \cdot (1 - 0.05)\%$ confidence interval for this mean weight as $(999.35, \ 1001.70)$. Now, we can put the problem differently and ask: "*is it true that the weight of a salt bag is 1000 g?*"

We may solve this problem by following the seven-step general procedure for hypothesis testing outlined in the previous section. This results in the statement of the null and alternative hypotheses:

$$H_0 : \ m = 1000$$
$$H_1 : \ m \neq 1000 \tag{2.5.2.1.5}$$

We set the significance level $\alpha = 0.05$. The calculated mean from the Sample Nr. 1 is $\bar{x} = 1000.53$, thus the value of the test statistic U is:

$$u = \frac{1000.53 - 1000}{3/\sqrt{25}} = 0.88 \tag{2.5.2.1.6}$$

From the tables of the standard normal distribution, the quantile $u(1 - 0.05/2)$ is:

$$u\left(1 - \frac{\alpha}{2}\right) = u\left(1 - \frac{0.05}{2}\right) = u(0.975) \approx 1.96 \tag{2.5.2.1.7}$$

Hence, the critical region is:

$$K_0 = (-\infty, -1.96) \cup (1.96, +\infty) \tag{2.5.2.1.8}$$

The calculated value of the test statistic does not fall into the critical region:

$$0.88 \notin (-\infty, -1.96) \cup (1.96, +\infty) \tag{2.5.2.1.9}$$

Therefore, we fail to reject the null hypothesis which states that the mean weight of the produced salt bags is 1000 g. Stated more completely, we conclude that there is no strong evidence that the mean salt bag's weight is different from 1000 g.

Below, we illustrate the above calculations using computer with R language.

R

We calculate the value of the test statistic:

```
> u=(1000.53-1000)/(3/sqrt(25))
[1] 0.88
```

We then read the value of the quantile $u(0.975)$ of the standard normal distribution:

```
> qnorm(0.975)
[1]  1.959964
```

and construct the critical region as in the above example, followed by the checking if this value falls into it.

Case 2. Standard Deviation σ is Unknown

The important result upon which this test procedure relies, is that the test statistic

$$T = \frac{\overline{X} - m_0}{S} \sqrt{n - 1} \tag{2.5.2.1.10}$$

where

$$S = \sqrt{\frac{1}{n} \sum_{i=1}^{n} (X_i - \overline{X})^2} \tag{2.5.2.1.11}$$

has t-distribution with $n - 1$ degrees of freedom if H_0 is true (this follows from Theorem 2.3.3 about the studentized sample mean). When we know the distribution of the test statistic if H_0 is true, we can locate the critical region to control the type I error probability at the desired level α. The structure of this test is identical to that for the case when σ is known, with the exception that the value σ is replaced by the computed value of the estimator S, and the standard normal distribution is replaced by the t-distribution when constructing the critical region:

$$K_0 = \left(-\infty, -t\left(1 - \tfrac{\alpha}{2}, n - 1\right)\right) \cup \left(t\left(1 - \tfrac{\alpha}{2}, n - 1\right), +\infty\right) \tag{2.5.2.1.12}$$

where $t(1 - \alpha/2, \ n - 1)$ is a quantile of order $1 - \alpha/2$ with $n - 1$ degrees of freedom of t-distribution.

Example 2.5.2.1.2 Hypothesis test for the population mean salt bag's weight (σ is unknown)

Let us consider the already investigated problem of the assessment of the mean weight of salt bags produced in a certain factory (Example 2.1.1, Sample 1). Assuming that the weight's distribution is approximately normal with the unknown standard deviation σ. As in Example 2.5.2.1.1, we set the significance level $\alpha = 0.05$.

We follow the seven-step general procedure for hypothesis testing. The hypothesis statement is the same as in Example 2.5.2.1.1. We calculate the value of the test statistic based on the Sample 1:

$$t = \frac{\overline{X} - m_0}{S}\sqrt{n - 1} = \frac{1000.53 - 1000}{2.27}\sqrt{25 - 1} = 1.138 \qquad (2.5.2.1.13)$$

The quantile of the t-distribution is:

$$t\left(1 - \frac{\alpha}{2}, n - 1\right) = t\left(1 - \frac{0.05}{2}, 25 - 1\right) = 2.064 \qquad (2.5.2.1.14)$$

which gives the critical region:

$$K_0 = (-\infty, -2.064) \cup (2.064, +\infty) \qquad (2.5.2.1.15)$$

As the calculated value of the test statistic does not fall into the critical region:

$$1.138 \notin (-\infty, -2.064) \cup (2.064, +\infty) \qquad (2.5.2.1.16)$$

we state that there is no strong evidence to reject the null hypothesis that the mean weight of a salt bag is 1000 g.

Case 3. Large Sample
In many if not most practical situations σ will be unknown. Furthermore, we may not be certain that the population is well modeled by a normal distribution. In these situations if n is large (**n > 30**) the procedure for the case of known σ can be easily converted into a large sample test procedure for unknown σ by simple substitution of the sample standard deviation S for σ in the test statistic

$$U = \frac{\overline{X} - m_0}{S/\sqrt{n}} \qquad (2.5.2.1.17)$$

Such procedure is valid regardless of the form of the distribution of the population. This large-sample test relies on the central limit theorem.

2.5.2.2 Hypothesis Tests on the Population Variance

Suppose that we wish to test the hypothesis that the variance σ^2 of a normal population equals a specified value, say σ_0^2. To test

$$H_0 : \ \sigma^2 = \sigma_0^2 \tag{2.5.2.2.1}$$

$$H_1 : \ \sigma^2 \neq \sigma_0^2 \tag{2.5.2.2.2}$$

we will use the test statistic

$$X^2 = \frac{n \cdot S^2}{\sigma_0^2} \tag{2.5.2.2.3}$$

which has χ^2 distribution with $n - 1$ degrees of freedom when H_0 is true (it follows from Theorem 2.3.5 about the sample variance).

Therefore, we calculate the critical region K_0 of the test statistic:

$$K_0 = \left(0, \ \chi^2\!\left(\tfrac{\alpha}{2}, n - 1\right)\right) \cup \left(\chi^2\!\left(1 - \tfrac{\alpha}{2}, n - 1\right), +\infty\right) \tag{2.5.2.2.4}$$

where $\chi^2\!\left(\tfrac{\alpha}{2}, n - 1\right)$ and $\chi^2\!\left(1 - \tfrac{\alpha}{2}, n - 1\right)$ are quantiles of the χ^2 distribution with $n - 1$ degrees of freedom of orders $\alpha/2$ and $(1 - \alpha/2)$, respectively.

Example 2.5.2.2.1 Let us consider the problem of the assessment of the variance of weight of salt bags produced in a certain factory (Example 2.1.1, Sample 1). As in Example 2.5.2.1.1, we set the significance level $\alpha = 0.05$.

We may solve this problem by following the seven-step general procedure for hypothesis testing. This results in the statement of the null and alternative hypotheses:

$$H_0 : \ \sigma^2 = 3^2$$
$$H_1 : \ \sigma^2 \neq 3^2 \tag{2.5.2.2.5}$$

Based on Sample Nr. 1, we calculate the value of the test statistic:

$$\chi_1^2 = \frac{n S^2}{\sigma_0^2} = \frac{25 \cdot (5, 13)^2}{3^2} = 14, 25 \tag{2.5.2.2.6}$$

For the assumed significance level $\alpha = 0.05$ the quantiles of χ^2 distribution of order $\alpha/2$ and $(1 - \alpha/2)$, respectively, are:

$$\chi^2\!\left(\frac{\alpha}{2}, n - 1\right) = \chi^2\!\left(\frac{0.05}{2}, 25 - 1\right) = 12, 40 \tag{2.5.2.2.7}$$

$$\chi^2\left(1 - \frac{\alpha}{2}, n - 1\right) = \chi^2\left(1 - \frac{0.05}{2}, 25 - 1\right) = 39,36 \qquad (2.5.2.2.8)$$

Hence, the critical region can be constructed as:

$$K_0 = (0, 12.40) \cup (39.36, +\infty) \qquad (2.5.2.2.9)$$

Since the value of the test statistic does not fall into the critical region:

$$14,25 \notin K_0 = (0, 12.40) \cup (39.36, +\infty) \qquad (2.5.2.2.10)$$

we find, there is no evidence for rejecting the null hypothesis H_0.

2.5.2.3 Hypothesis Tests for a Difference in Two Means

The previous section presented statistical inference based on hypothesis tests for a single population parameter (the mean m, the variance σ^2). This and the next section extend those results to the case of two independent normal populations with distributions: $N(x; m_1, \sigma_1)$ and $N(x; m_2, \sigma_2)$. Inferences will be based on the two independent random samples of sizes n_1 and n_2, respectively. We now consider the hypothesis testing on the difference in the means

$$H_0 : m_1 = m_2$$
$$H_1 : m_1 \neq m_2 \qquad (2.5.2.3.1)$$

We will present two cases here.

Case 1 We will assume that standard deviations σ_1, σ_2 are known. From Theorem 2.3.2, it follows that the statistic:

$$U_2 = \frac{\overline{X}_1 - \overline{X}_2 - (m_1 - m_2)}{\sqrt{\frac{\sigma_1^2}{n_1} + \frac{\sigma_2^2}{n_2}}} \qquad (2.5.2.3.2)$$

follows the standard normal distribution $N(x; 0, 1)$ if the null hypothesis is true. Assuming the truth of the null hypothesis $((m_1 - m_2) = 0)$, the test statistic:

$$U = \frac{\overline{X}_1 - \overline{X}_2}{\sqrt{\frac{\sigma_1^2}{n_1} + \frac{\sigma_2^2}{n_2}}} \qquad (2.5.2.3.3)$$

also follows standard normal distribution. Based on this fact, we can construct two-tailed critical region of this test statistic on the specified level of significance α.

$$K_0 = \left(-\infty, -u\left(1 - \frac{\alpha}{2}\right)\right) \cup \left(u\left(1 - \frac{\alpha}{2}\right), +\infty\right) \qquad (2.5.2.3.4)$$

Case 2 We will now consider the case when the standard deviations σ_1, σ_2 are unknown, but equal $\sigma_1 = \sigma_2 = \sigma$. In most situations we do not know whether this assumption is met, therefore, first we must verify the hypothesis about equality of variances (presented in the next section).

The test statistic in this case has the form:

$$T = \frac{\overline{X}_1 - \overline{X}_2}{\sqrt{\frac{n_1 \cdot S_1^2 + n_2 \cdot S_2^2}{n_1 + n_2 - 2}\left(\frac{1}{n_1} + \frac{1}{n_2}\right)}} \tag{2.5.2.3.5}$$

where

$$S_1^2 = \frac{1}{n_1}\sum_{i=1}^{n_1}\left(X_i - \overline{X}_1\right)^2 \quad S_2^2 = \frac{1}{n_2}\sum_{i=1}^{n_2}\left(X_i - \overline{X}_2\right)^2 \tag{2.5.2.3.6}$$

has t distribution with $n_1 + n_2 - 2$ degrees of freedom which follows from Theorem 2.3.4. This fact allows to construct the critical region:

$$K_0 = \left(-\infty, -t\left(1 - \frac{\alpha}{2}, n_1 + n_2 - 2\right)\right) \cup$$
$$\left(t\left(1 - \frac{\alpha}{2}, n_1 + n_2 - 2\right), +\infty\right) \tag{2.5.2.3.7}$$

Example 2.5.2.3.1 Two sample test for comparing population means
Let us now consider the problem of comparing the mean weights of salt bags produced in the two different factories. Let us assume that the weights distributions are normal $N(m_1, \sigma_1 = 3)$ and $N(m_2, \sigma_2 = 5)$. Two independent samples of $n_1 = n_2 = 25$ bags have been taken from each company.

The sample from the first company is the already known **Sample No. 1**, while the **Sample No. 2** consists of the following 25 observations:

> 1001.08, 1004.89, 991.86, 999.48, 1002.54, 996.83, 997.41, 1003.30, 1005.05, 999.14, 989.04, 1003.13, 994.79, 1003.32, 996.33, 990.90, 992.84, 999.24, 1005.54, 1001.13, 996.56, 1004.00, 995.09, 1006.98, 1005.03

We test the null hypothesis:

$$H_0 : m_1 = m_2 \tag{2.5.2.3.8}$$

at the significance level $\alpha = 0.05$. The value of the test statistic is:

$$u = \frac{\overline{X}_1 - \overline{X}_2}{\sqrt{\frac{\sigma_1^2}{n_1} + \frac{\sigma_2^2}{n_2}}} = \frac{1000.53 - 999.42}{\sqrt{\frac{3^2}{25} + \frac{5^2}{25}}} = 0.95 \tag{2.5.2.3.9}$$

The critical region:

$$K_0 = (-\infty, -1.96) \cup (1.96, +\infty) \tag{2.5.2.3.10}$$

As the value of the test statistic does not fall into the critical region, we have no strong evidence to reject the null hypothesis about the equalities of population weights of salt bags produced in the two factories.

2.5.2.4 Two Sample Test Concerning Variances

Now, let us consider the problem of testing the equality of the two variances σ_1^2, σ_2^2 of the two normal populations: $N(x; m_1, \sigma_1)$ and $N(x; m_2, \sigma_2)$, respectively. Assume that two independent samples of sizes n_1 and n_2 have been taken from the two populations. We wish to test the following null hypothesis against the alternative:

$$H_0 : \sigma_1^2 = \sigma_2^2$$
$$H_1 : \sigma_1^2 \neq \sigma_2^2 \tag{2.5.2.4.1}$$

From Theorem 2.3.6, it follows that the test statistic:

$$F = \frac{S_{*1}^2}{S_{*2}^2} \tag{2.5.2.4.2}$$

where

$$S_{*1}^2 = \frac{n_1 S_1^2}{n_1 - 1} \quad S_{*2}^2 = \frac{n_2 S_2^2}{n_2 - 1} \tag{2.5.2.4.3}$$

are unbiased estimators of variances in the first and the second population, respectively, has F-distribution with $n_1 - 1$ and $n_2 - 1$ degrees of freedom. Based on this fact, we can construct, on the prespecified significance level α, the critical region of the above statistic:

$$K_0 = (f(1 - \alpha, n_1 - 1, n_2 - 1), +\infty) \tag{2.5.2.4.4}$$

Example 2.5.2.4.1 Two sample test for comparing population variances

Let us now consider the problem of comparing the two variances of distribution of salt bag's weights produced in the two different factories, except that the variances are not known.

At the significance level $\alpha = 0.05$, we verify the null hypothesis:

$$H_0 : \sigma_1^2 = \sigma_2^2 \tag{2.5.2.4.5}$$

Based on the samples, the calculated values of unbiased sample variances are:

$$s_{*1}^2 = 5.34 \quad s_{*2}^2 = 25.74 \tag{2.5.2.4.6}$$

The value of the test statistic is:

$$f = \frac{s_{*1}^2}{s_{*2}^2} = 4.82 \tag{2.5.2.4.7}$$

And the critical region, based on the quantile $f(1 - 0.05, 24, 24) = 1.98$ of F-distribution is:

$$K_0 = (1.98, +\infty) \tag{2.5.2.4.8}$$

The value of the test statistic falls into the critical region, therefore, at the significance level $\alpha = 0.05$, we reject the null hypothesis about the equality of variances of salt bag's weights in the two factories.

2.5.3 Nonparametric Tests

In the previous sections we have presented the parametric tests, that means the hypothesis-testing procedures for problems in which the probability distribution is known, and the hypotheses involve the parameters of the distribution. Often, we do not know the underlying distribution of the population, and we wish to test the hypothesis that a particular distribution will be satisfactory as a population model. For example, we might wish to test the hypothesis that the population is normal. Such tests are the examples of *nonparametric tests* and are called the *goodness of fit tests*. We will describe three examples of such tests: the test based on a chi-square distribution, the Kolmogorov test and Shapiro-Wilk's test. Moreover, we present the *test for independence* of two variables, also based on a chi-square distribution.

Generally, *nonparametric tests*, as being a part of nonparametric statistics, are distribution free methods which do not rely on assumptions that the data are drawn from a given parametric family of probability distributions. If the assumptions for the parametric method may be difficult, or impossible to justify, or if the data are reported on nominal scale, then nonparametric tests should be used. For example, if the sample mean of the two populations being compared does not follow a normal distribution, the already presented popular and commonly used *t*-test for comparing two population means should not be used. The nonparametric test, *the Mann–Whitney U test* also known as *Wilcoxon rank-sum test* is a good alternative and will also be described in this section.

The seven-step general hypothesis-testing procedure described earlier, can also be used now, in nonparametric settings.

2.5.3.1   Chi-Square Goodness of Fit Test

Chi-Square goodness of fit test is a non-parametric test that is used to find out how the observed values of a given phenomena are significantly different from the expected values. In this test, the term *goodness of fit* is used to compare the observed sample distribution with the expected probability distribution. Chi-Square goodness of fit test determines how well theoretical distribution (such as normal, binomial, Poisson, etc.) fits the empirical distribution.

The null hypothesis H_0 in this test assumes that there is no significant difference between the observed and the expected values of the hypothesized distribution of random variable X. The alternative hypothesis H_1 states that there is a significant difference between them. The test procedure requires a random sample of size n from the population whose probability distribution is unknown. These n observations are arranged in a frequency table having k classes: points or intervals depending on the discrete or continuous nature of the data, respectively (see Table 2.5). The value n_i in that table is called the *observed frequency* of data in ith class. From the hypothesized probability distribution, we compute then the *expected frequency* in the ith class, in the following way. First, we compute the theoretical, hypothesized probability associated with the ith class: in the case of point frequency table as:

$$p_i = P_0(X = x_i) \tag{2.5.3.1.1}$$

while for interval-based one, as:

$$p_i = P_0\{x_{0i} < X \le x_{1i}\} = F_0(x_{1i}) - F_0(x_{0i}) \tag{2.5.3.1.2}$$

where $F_0(x)$ denotes the cumulative distribution function of X. The expected frequencies are then computed by multiplying the sample size n by the probabilities p_i. The test statistic is:

$$\chi^2 = \sum_{i=1}^{k} \frac{(n_i - np_i)^2}{np_i} \tag{2.5.3.1.3}$$

Table 2.5 Frequency table and calculation of X^2 statistic (Example 2.5.3.1.1)

i	n_i	np_i	$(n_i - np_i)^2$	$(n_i - np_i)^2/np_i$
1	11	20	81	4.05
2	30	20	100	5.0
3	14	20	36	1.80
4	10	20	100	5.0
5	33	20	169	8.45
6	22	20	4	0.20
				24.50

Assuming the truth of the null hypothesis, the above test statistic has approximately $\chi^2(k-1)$ distribution, or $\chi^2(k-m-1)$, if m unknown parameters of hypothesized distribution are estimated by the maximum likelihood method. Based on this fact, we can construct, for a given level of significance α, the critical region:

$$\left(\chi^2(1-\alpha, k-m-1), +\infty\right) \qquad (2.5.3.1.4)$$

where $\chi^2(1-\alpha, k-m-1)$ is a quantile of order $1-\alpha$ of chi-square distribution with $(k-m-1)$ degrees of freedom.

One point to be noted in the application of this test procedure concerns the magnitude of the expected frequencies. There is no general agreement regarding the minimum value of expected frequencies, but values of 3, 4, and 5 are widely used as minimal.

Example 2.5.3.1.1 We wish to verify the reliability of the purchased die. In our null hypothesis, we hypothesize that the distribution is (discrete) uniform, that is:

$$p_i = P_0(X = i) = \frac{1}{6} \quad \text{for } i = 1, \ldots, 6 \qquad (2.5.3.1.5)$$

To verify this hypothesis the chi-squared goodness of fit test, $n = 120$ rolls were made, the obtained results are presented in the first two columns of point frequency Table 2.5.

The 3rd column in Table 2.5 comprises the expected frequencies, while the last two—the calculation of the value of test statistic, which is $\chi^2 = 24.50$. The appropriate quantile of chi-square distribution is $\chi^2(1-0.05, 5) = 11.070$. As $24.50 > 11.07$, the value of test statistic falls into the critical region, we reject the null hypothesis at the significance level $\alpha = 0.05$. There is strong evidence that the purchased die is not reliable.

2.5.3.2 Chi-Square Test for Independence

The chi-square test for independence is used to determine whether there is a significant relationship between the two nominal (categorical) variables. Suppose we wish to examine whether the company productivity depends on the level of absence of their workers. A sample of n companies is randomly selected, then the productivity and the level of absence are measured on r- and k-point scales, respectively. The data, i.e. the *observed frequencies* n_{ij} can then be displayed in the *contingency table* (see Table 2.6), where each of r rows represents a category of the first variable, and each of the k columns represents a category of the second variable. The elements of last row and last column contain the sums of corresponding column or row observed frequencies, respectively,

Table 2.6 The contingency table for chi-squared test of independence

X	Y				
	y_1	y_2	$\cdots$	y_k	$n_{i.} = \sum\limits_{j=1}^{k} n_{ij}$
x_1	n_{11}	n_{12}	$\cdots$	n_{1k}	$n_{1.}$
x_2	n_{21}	n_{22}	$\cdots$	n_{2k}	$n_{2.}$
$\cdots$	$\cdots$	$\cdots$	$\cdots$	$\cdots$	$\cdots$
x_r	n_{r1}	n_{r2}	$\cdots$	n_{rk}	$n_{r.}$
$n_{.j} = \sum\limits_{i=1}^{r} n_{ij}$	$n_{.1}$	$n_{.2}$	$\cdots$	$n_{.k}$	$\sum\limits_{i=1}^{r}\sum\limits_{j=1}^{k} n_{ij} = 1$

$$n_{i.} = \sum_{j=1}^{r} n_{ij} \; i = 1,\ldots,r \quad n_{.j} = \sum_{i=1}^{k} n_{ij} \; i = 1,\ldots,k \qquad (2.5.3.2.1)$$

and are called *marginal frequencies*.

Of course

$$n = \sum_{i=1}^{k} n_{i.} = \sum_{j=1}^{r} n_{.j} = \sum_{i=1}^{k}\sum_{j=1}^{r} n_{ij} \qquad (2.5.3.2.2)$$

Our decision to accept, or reject the null hypothesis H_0 of independence between company productivity, and the level of absence, is based upon how good a fit we have between the observed frequencies in each of the $k \times r$ cells of contingency table, and the frequencies that we would expect for each cell under the assumption that H_0 is true.

To find these expected frequencies, let us consider the imagined population characterized by the two discrete random variables X and Y. The joint probability distribution of random vector (X, Y) can be described by the two-dimensional table (similar to that in Sect. 1.5):

$$\left(p_{ij} = P(X = x_i, Y = y_j) : \; i = 1,\ldots,r; \; y = 1,\ldots,k\right) \qquad (2.5.3.2.3)$$

The *marginal probabilities* are:

$$p_{i.} = \sum_{j=1}^{k} p_{ij} \quad p_{.j} = \sum_{i=1}^{r} p_{ij} \qquad (2.5.3.2.4)$$

The null hypothesis of independence can be rewrite as:

$$H_0 : \; \forall \, i, j \quad p_{ij} = p_{i.} \cdot p_{.j} \qquad (2.5.3.2.5)$$

which relies on the definition of stochastic independence of random variables X and Y. The maximum likelihood estimators of marginal probabilities are:

$$\hat{p}_{i.} = \frac{n_{i.}}{n} \quad \hat{p}_{.j} = \frac{n_{.j}}{n} \tag{2.5.3.2.6}$$

Given the null hypothesis of independence is true, it is expected that the number in the cell (i, j) is:

$$n \cdot p_{ij} = n \cdot \hat{p}_{i.} \cdot \hat{p}_{.j} = \frac{n_{i.} \cdot n_{.j}}{n} \tag{2.5.3.2.7}$$

This quantity

$$\hat{n}_{ij} = \frac{n_{i.} \cdot n_{.j}}{n} \tag{2.5.3.2.8}$$

is called the *expected frequency*.

Large differences between the observed n_{ij} and the expected $\hat{n}_{ij}$ frequencies, respectively, provide strong evidence against the null hypothesis, and has been used for the construction of the *test statistic* in the *chi-square test of independence*:

$$\chi^2 = \sum_{i=1}^{k} \sum_{j=1}^{r} \frac{(n_{ij} - \hat{n}_{ij})^2}{\hat{n}_{ij}} \tag{2.5.3.2.9}$$

Assuming the null hypothesis is true, the above test statistic has asymptotic chi-squared distribution with $(r - 1)(k - 1)$ degrees of freedom.

For a given level of significance α, the critical region for this test is:

$$\left(\chi^2(1 - \alpha, (r - 1)(k - 1)), +\infty \right) \tag{2.5.3.2.10}$$

where $\chi^2(1 - \alpha, (r - 1)(k - 1))$ is a quantile of order $(1 - \alpha)$ of chi-square distribution with $(r - 1)(k - 1)$ degrees of freedom. The rejection of null hypothesis in the test of independence means the relationship between the variables X and Y is statistically significant.

We may be interested in the strength of the relationship between the two variables. This can be measured by the *V Cramer's coefficient*:

$$V = \sqrt{\frac{\chi^2}{n \cdot \min(k, r)}} \tag{2.5.3.2.11}$$

where χ^2 is the value of the test statistic. This coefficient takes values from the interval $[0, 1]$, the value 0 when the variables are independent, the value 1 in the case of the perfect functional dependence.

Table 2.7 Contingency table

	Prod.-small	Prod.-medium	Prod.-large	$n_{i.}$
Abs-small	9	5	3	17
Abs-medium	4	12	6	22
Abs-large	1	6	14	21
$n_{.j}$	14	23	23	60

Example 2.5.3.2.1 In order to examine whether the company productivity depends on the level of absence of their workers, a sample of 60 companies was randomly selected, then the productivity and the level of absence were measured on a 3-point scale (small/medium/large) in each of the company. The results are presented in the contingency table (Table 2.7, the calculated marginal counts are also included).

Lets' verify the hypothesis about the independence of the company productivity and the level of absence at the significance level $\alpha = 0.05$.

We calculate the expected frequencies:

$$\hat{n}_{11} = \frac{14 \cdot 17}{60} = 3.97, \hat{n}_{12} = \frac{17 \cdot 23}{60} = 6.52, \ldots \hat{n}_{33} = \frac{21 \cdot 23}{60} = 8.05$$

Then, the value of the test statistic:

$$\lambda^2 = \frac{(9 - 3.97)^2}{3.97} + \frac{(5 - 6.52)^2}{6.52} + \cdots + \frac{(14 - 8.05)^2}{8.05} = 19.2$$

Based on the quantile $\chi^2(1 - 0.05, 4) = 9.49$, we then construct the critical region $(9.49, +\infty)$. Since the calculated value of the test statistic falls into the critical region, we reject the null hypothesis on the independence of company's productivity and the level of absence at the significance level $\alpha = 0.05$.

The strength of this relationship can be quantified by the value of V-Cramer coefficient:

$$V = \sqrt{\frac{19.2}{60 \cdot 3}} \approx 0.3265$$

2.5.3.3 Kolmogorov's Goodness of Fit Test

The Kolmogorov's goodness of fit test checks how well theoretical distribution fits the empirical distribution. It is based on test statistic:

$$D_n = \max_{x} |F_0(x) - \hat{F}_n(x)| \tag{2.5.3.3.1}$$

where $\hat{F}_n(x)$ is the empirical cumulative distribution function of a sample $(x_1, \ldots, x_n)$, and $F_0(x)$ is the theoretical (hypothetical) cumulative distribution function.

It can be shown that the value of the D_n test statistic is equal to:

$$D_n = \max(d_n^-, d_n^+) \tag{2.5.3.3.2}$$

where

$$d_n^+ = \max_{1 \le i \le n} \left| \frac{i}{n} - F_0(x) \right| \tag{2.5.3.3.3}$$

$$d_n^- = \max_{1 \le i \le n} \left| F_0(x) - \frac{i-1}{n} \right| \tag{2.5.3.3.4}$$

Moreover, it can be shown that if $F_0(x)$ is continuous, then a distribution of D_n test statistic does not depend on the form of $F_0(x)$, and it is known as the *exact Kolmogorov's distribution*. The critical region of this test is:

$$K_0 = \langle d_n(1-\alpha), \; 1 \rangle \tag{2.5.3.3.5}$$

where the critical value $d_n(1-\alpha)$ can be found in the tables of the exact Kolmogorov's distribution for appropriate values of n and α.

If the sample size n is large ($n \le 100$), like a few hundreds, then the *limit distribution of* D_n statistic can be used. This fact results from the following theorem.

Theorem 2.5.3.3.1 *If* $n \to \infty$, *then:*

$$P(\sqrt{n}D_n \ge d) \to K(d) = \sum_{i=-\infty}^{i=+\infty} (-1)^i e^{-2i^2 d^2} \quad (d > 0) \tag{2.5.3.3.6}$$

From the above theorem, it follows that the limit $K(d)$ is a known cumulative distribution function. Thus, for a given significance level α, the $d(1 - \alpha)$ quantiles of the limit Kolmogorov's distribution can be found.

If the value $\sqrt{n} \cdot D_n$ calculated for a given sample is greater than $d(1 - \alpha)$, then at the level of significance α, we reject the verified hypothesis.

Example 2.5.3.3.1 We wish to verify at the significance level $\alpha = 0.05$ that the 25-element sample of salt bag's weights (**Sample No. 1**, Example 2.1.1) comes from a normal population.

From the sample, we calculate the values of estimates of the two parameters in normal distribution:

$$\hat{m} = \frac{1}{25} \sum_{i=1}^{25} x_i = 1000.53 \tag{2.5.3.3.7}$$

$$\hat{\sigma} = \sqrt{\frac{1}{25} \sum_{i=1}^{25} (x_i - \hat{m})^2} = 2.27 \tag{2.5.3.3.8}$$

Calculations of Kolmogorov's statistic are shown in Table 2.8.

For the standardized x values, i.e. $(x_i - \hat{m})/\hat{\sigma}$, the values of the corresponding cumulative distribution function $F_0(x)$ were taken from tables of standard normal distribution. The table shows that $d_{25}^+ = 0.08$, $d_{25}^- = 0.09$ resulting in $D_{25} = \max(d_{25}^-, d_{25}^+) = 0.09$. The critical value of the distribution of D_{25}-statistics read from Kolmogorov's distribution tables for the assumed level of significance $\alpha = 0.05$ is $d_{25}(1-0.05) = 0.264$. Because $0.09 \notin \langle 0.264, \ 1\rangle$, so the Kolmogorov's

Table 2.8 Calculations of Kolmogorov's test statistic (Example 2.5.3.3.1)

i	x	Standard. x	i/n	F0(x)	|i/n − F0(x)|	|(i − 1)/n − F0(x)|
1	995.83	−2.08	0.04	0.02	0.02	0.00
2	996.25	−1.89	0.08	0.03	0.05	0.01
3	997.64	−1.27	0.12	0.10	0.02	0.02
4	998.01	−1.11	0.16	0.13	0.03	0.01
5	998.09	−1.08	0.2	0.14	0.06	0.02
6	998.89	−0.72	0.24	0.23	0.01	0.03
7	998.98	−0.68	0.28	0.25	0.03	0.01
8	999.56	−0.43	0.32	0.33	0.01	0.05
9	999.56	−0.43	0.36	0.34	0.02	0.02
10	999.58	−0.42	0.4	0.34	0.06	0.02
11	1000.33	−0.09	0.44	0.47	0.03	0.07
12	1000.56	0.01	0.48	0.51	0.03	0.07
13	1000.85	0.14	0.52	0.56	0.04	0.08
14	1001.16	0.28	0.56	0.61	0.05	0.09
15	1001.42	0.39	0.6	0.65	0.05	0.09
16	1001.59	0.47	0.64	0.68	0.04	0.08
17	1001.68	0.51	0.68	0.69	0.01	0.05
18	1001.72	0.53	0.72	0.70	0.02	0.02
19	1001.80	0.56	0.76	0.71	0.05	0.01
20	1002.08	0.69	0.8	0.75	0.05	0.01
21	1002.14	0.71	0.84	0.76	0.08	0.04
22	1002.80	1.00	0.88	0.84	0.04	0.00
23	1003.26	1.21	0.92	0.89	0.03	0.01
24	1004.42	1.72	0.96	0.96	0.00	0.04
25	1004.97	1.96	1	0.98	0.02	0.02
				Max	0.08	0.09

test does not reject the null hypothesis that the population distribution of salt bag's weight is a normal distribution.

2.5.3.4 Shapiro-Wilk Test of Normality

Normality is one of the most common assumptions when using different statistical procedures. There are a lot of tests for checking normality, including the Kolmogorov test described in the previous section. All normality tests are sensitive to sample size. If it is lower than 50 observations, that it is preferable to use the described below Shapiro Wilk test. Also graphical methods are a good alternative to evaluate normality, in particular QQ (*quantile-quantile*) plots, by plotting data quantiles against those of standard normal distribution.

The Shapiro-Wilk test is used to verify the null hypothesis that the distribution of population (random variable X) is normal. It is based on test statistic:

$$W = \frac{\left[\sum_{i=1}^{n} a_i X_{(i)}\right]^2}{\sum_{i=1}^{n} \left(X_i - \overline{X}\right)^2} \qquad (2.5.3.4.1)$$

$X_{(1)} \leq X_{(2)} \leq \ldots \leq X_{(n)}$ are the ordered values of the sample $(X_1, \ldots, X_n)$, a_i are tabulated coefficients. Small values of W indicate non-normality. The null hypothesis is rejected at the significance level α if:

$$W \leq W(\alpha, n) \qquad (2.5.3.4.2)$$

where $W(\alpha, n)$ are the critical values of the distribution W.

2.5.3.5 Mann-Whitney U Test

When we are interested in testing equality of means of two continuous distributions that are obviously non-normal, and samples are independent, the nonparametric test, the *Mann-Whitney U* test (also known as *Wilcoxon rank-sum test*), is the appropriate alternative to the two-sample t-test for comparing population means, described earlier.

We will develop the *Mann-Whitney* test in a specific context. Suppose that we have $n_1 + n_2$ experimental units to assign to a treatment group and a control group. The assignment is made at random: n_1 units are randomly chosen and assigned to the control, and the remaining n_2 units are assigned to the treatment. We are interested in testing the null hypothesis that the treatment has no effect. If the null hypothesis is true, then any difference in the outcomes is due to the randomization.

A test statistic is calculated in the following way. First, we arrange the $n_1 + n_2$ observations of the combined samples in ascending order and substitute a rank of $1, 2, \ldots, n_1 + n_2$ for each observation. In the case of ties (identical observations), we replace the observations by the mean of the ranks that the observations would have if they were distinguishable. The sum of the ranks corresponding to the n_1 observations in the smaller sample is denoted by R_1. Similarly, the value R_2 represents the sum of the n_2 ranks corresponding to the larger sample.

The total $R_1 + R_2$ depends only on the number of observations in the two samples $(R_1 + R_2 = (n_1 + n_2)(n_1 + n_2 + 1)/2)$. The null hypothesis will be rejected if R_1 is small and R_2 is large or vice versa. In actual practice, we usually base our decision on the value of test statistic:

$$U = \min(U_1, \ U_2) \tag{2.5.3.5.1}$$

$$U_1 = R_1 - \frac{n_1(n_1 + 1)}{2} \quad U_2 = R_2 - \frac{n_2(n_2 + 1)}{2} \tag{2.5.3.5.2}$$

The null hypothesis is rejected at the significance level α, when the value of the U statistic calculated from the sample is less than or equal to the tabled critical value.

Example 2.5.3.5.1 The following data show the number of defects in some products manufactured by the two methods of which the second is the improved version of the first:

$$X_1 : 30\ 32\ 20\ 23\ 44\ 31\ 28\ 33$$
$$X_2 : 41\ 46\ 29\ 32\ 20\ 23\ 48\ 36\ 42\ 43$$

We wish to know if there a difference in the number of defects before, and after the modification, in other words, we ask "did the modification introduce an **effect**?"

The observations are arranged in ascending order and ranks are assigned to them (3rd row):

20 20 23 23 28 29 30 31 32 32 33 36 41 42 43 44 46 48
1   2   1   2   1   2   1   1   1   2   1   2   2   2   2   1   2   2
1.5 1.5 3.5 3.5 5   6   7   8 9.5 9.5 11 12 13 14 15 16 17 18

In the second row, the labels of population have been preserved for clarity. We calculate the sum of ranks for the first population:

$$R_1 = 1.5 + 3.5 + 5 + 7 + 8 + 9.5 + 11 + 16 = 61.5 \tag{2.5.3.5.3}$$

Similarly, $R_2 = 109.5$ and then the value of test statistics:

$$u = \min(61.5;\ 109.5) = 61.5 \tag{2.5.3.5.4}$$

The critical value for $n_1 = 8$, $n_2 = 10$ and the significance level $\alpha = 0.05$ is equal to u* = *17*.

Since, $u = 74.5 > 17$, at the significance level 0.05, we do not reject the null hypothesis. There is no significant difference in the number of defects before and after the modification of the technology (no effect).

2.5.4 P-Values in Hypothesis Tests

One way to present the results of a hypothesis test is to state that the null hypothesis H_0 is, or is not rejected at a specified significance level α. This statement is often inadequate because it gives the decision maker no idea about whether the computed value of the test statistic was just barely, or, far in the critical region. Moreover, reporting the results this way imposes the predefined level of significance α on other users of the information. This approach may be unsatisfactory because it might be difficult for some decision makers to have an idea about the risks implied by α!

To avoid these difficulties the p-value approach has been proposed, and now commonly used in practice. *The p-value is the probability that the test statistic will take on a value that is at least as extreme as the observed value of the statistic, when the null hypothesis H_0 is true.* More formally, *p-value* is the smallest level of significance that would lead to rejection of the null hypothesis H_0 with the given data.

It is important to notice that p-value conveys much information about the weight of evidence against H_0, and allows a decision maker to draw a conclusion at any specified level of significance. Having computed p-value based on a given data, we know that the null hypothesis H_0 would be rejected at any level of significance $\alpha \geq p$-value. Thus, the p-value associated with a specific decision gives us the opportunity for displaying the risk levels.

On the other hand, the purpose is most often only to decide whether to reject or not the null hypothesis H_0 at a given level of significance α. In such cases, the approach to hypothesis testing based on p-values leads to the commonly used rule for making decision:

- we reject the null hypothesis H_0 when:

$$p\text{-value} \leq \alpha \tag{2.5.4.1}$$

- there is no evidence to reject the null hypothesis when:

$$p\text{-}value > \alpha \tag{2.5.4.2}$$

If the p-value approach is used for hypothesis testing, it is not necessary to state explicitly the critical region in the presented, general procedure for hypothesis testing.

Most modern computer programs for statistical analysis report p-values. Below, we illustrate the approach to hypothesis testing based on p-values using R. A short introduction to calculations in R is in Appendix B.

Example 2.5.4.1 Hypothesis test for the population mean salt bag's weight based on *p*-value

Let us consider the already investigated problem of the assessment of the mean weight of salt bags produced in a certain factory (Example 2.1.1, Sample Nr. 1). Our null hypothesis is $H_0 : m = 1000 \, g$.

R

The values of salt bag's weights forming a random sample are combined into a vector and assigned variable *data*:

```
> data = c(1000.33, 1004.97, 998.98, 1000.85, 1001.42,
1001.68, 999.58,    1001.16, 1001.79,  997.64,  1001.59,
1000.56, 1003.26, 996.25, 995.83, 999.56, 1002.08, 998.89,
998.09, 1004.42, 1002.14, 998.01, 1002.79, 999.56,
1001.72)
```

On the created variable *data*, further calculations are carried out. We first check the assumption—verify whether the data comes from a normal distribution:

```
> ks.test(data,"pnorm",mean(data),sd(dane))
One-sample Kolmogorov-Smirnov test
                data:   data
                D = 0.0906, p-value = 0.9865
                alternative hypothesis: two-sided
```

Because the obtained *p*-value $= 0.9865 > 0.05$, there is no evidence to reject the null hypothesis about the normality of weight distribution.

Then, we perform the test for the mean weight by calling the function *t.test* with the appropriate parameters:

```
> t.test(data, mu=1000)
                One Sample t-test
                            data:   data
t = 1.1383, df = 24, p-value = 0.2662
```

Because:

$$p\text{-}value = 0.2662 > \alpha = 0.05$$

there is no evidence to reject the null hypothesis.

Example 2.5.4.2 This illustrative example presents the testing procedure for the comparison of two means on the simulated data (the case of unknown variances described previously). The data are generated from the two different normal distributions.

R

```
> x=rnorm(50,mean=5,sd=1)
> y=rnorm(30,mean=7,sd=1)
```

Assuming (for a moment) that we do not know standard deviation of weight distributions, we first verify the hypothesis about equality of variances in two populations:

```
> var.test(x,y)
          F test to compare two variances
data:   x and y
F = 1.3324, num df = 49, denom df = 29, p-value = 0.4121
alternative hypothesis: true ratio of variances is not equal to 1
95 percent confidence interval:
 0.6694131 2.5067412
sample estimates:
ratio of variances
          1.332369
```

Because p-value $= 0.4121 > 0.05$, there is no evidence for rejecting the null hypothesis. So, we can perform a test for the comparison of two means.

```
> t.test(x,y)
         Welch Two Sample t-test
data:   x and y
t = -9.0594, df = 68.131, p-value = 2.595e-13
alternative hypothesis: true difference in means is not equal to 0
95 percent confidence interval:
 -2.497739 -1.596047
sample estimates:
mean of x mean of y
 5.042800  7.089693
```

Due to the very low p-value, less than $\alpha = 0.05$, we reject the null hypothesis about the equality of means in the two considered populations.

2.5.5 Permutation Tests

As we saw in the previous sections, standard parametric statistical tests have numerous assumptions that should be respected prior to drawing conclusions from their results. Violation of these assumptions can lead to erroneous conclusions about the populations under study. Permutation tests provide yet another alternative solution to parametric tests.

A *permutation test* (also called a *randomization test*) is a type of statistical significance test in which the distribution of the test statistic under the null hypothesis is obtained by calculating all possible values of the test statistic under rearrangements of the labels on the observed data points. Like bootstrapping, a permutation test builds sampling distribution (called the "*permutation distribution*") by resampling (without replacement) the observed data: it permutes the observed data by assigning different outcome values to each observation from among the set of actually observed outcomes. Permutation tests exist for any test statistic, regardless of whether or not its distribution is known. However, an important assumption behind a permutation test is that the observations are exchangeable under the null hypothesis. An important consequence of this assumption is that tests of difference in location (like a permutation t-test) require equal variances.

Permutation tests are particularly relevant in experimental studies, where we are often interested in the null hypothesis of no difference between two groups. We illustrate the basic idea of a permutation test through the following example.

Example 2.5.5.1 Consider the problem of comparing the mean weights of salt bags produced by the two different factories, say $f1$ and $f2$. Suppose the random variables X_1, X_2 represent the two weights of salt bags produced by the two considered factories, respectively. Let $n_1 = 25$ and $n_2 = 25$ be the sample sizes collected from each population. The sample means of weights in the two populations are:

$$\bar{x}_1 = 1000.53 \quad \bar{x}_2 = 999.42 \tag{2.5.5.1}$$

which gives the *observed difference*:

$$r_0 = \bar{x}_1 - \bar{x}_2 = 1.11 \tag{2.5.5.2}$$

We shall determine whether the observed difference between the sample means is large enough to reject, at some significance level α, the null hypothesis

$$H_0 : \ m_1 = m_2 \tag{2.5.5.3}$$

that there is no difference in mean salt bag's weights $E(X_1) = m_1$ and $E(X_2) = m_2$ of the two populations (i.e. that they come from the same distribution). A reasonable test statistic is:

$$R = \overline{X}_1 - \overline{X}_2 \tag{2.5.5.4}$$

The distribution of test statistics is not known, since no assumptions about the distributions of weights has been done.

The test proceeds as follows. First, the sample observations of two populations are pooled and the difference in sample means is calculated and recorded for every possible way of dividing the pooled values into the two groups of size n_1 and n_2 (for every permutation of the two group labels $f1$ and $f2$). With a different allocation of sample observations, a different value for R statistics would be obtained. The set of all calculated differences is the permutation distribution of possible differences (for this sample) under the null hypothesis.

While a permutation test requires that we see all possible permutations of the data (which can become quite large), we can easily conduct "*approximate permutation tests*" by generating the reference distribution by Monte Carlo sampling, which takes a small (relative to the total number of permutations) random sample of the possible replicates, in our example $B = 1000$. These 1000 permutations give the histogram shown in Fig. 2.12 which approximates the unknown distribution of R statistics.

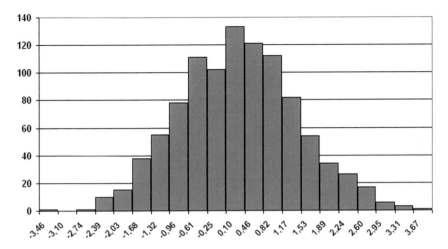

Fig. 2.12 Histogram of the distribution of R statistics for $B = 1000$ random permutations

The one-sided p-value of the test is calculated as the proportion of sampled permutations where the difference in means was greater than or equal to the observed one:

$$p = \frac{\#\{|r_i| \geq r_0 \quad i = 1, \ldots, B\}}{B} \tag{2.5.5.5}$$

where r_i is the value of the test statistic in the ith permutation. In our example, we obtained $p = 0.332$, which gives an approximate probability of obtaining the value $r_0 = 1.11$ and higher when the null hypothesis is true.

Hence, $0.332 > \alpha = 0.05$, there is no strong evidence for the rejection of the null hypothesis, that means for the difference in the two mean salt bag's weights.

Bibliography

Hodges Jr., J.L., Lehmann, E.L.: Basic Concepts of Probability and Statistics, 2nd edn. Society for Industrial and Applied Mathematics, Philadelphia (2005)

R Core Team: R language definition. http://cran.r-project.org/doc/manuals/r-release/R-lang.pdf (2019). Accessed 16 Dec 2019

Rice, J.: Mathematical Statistics and Data Analysis, 3rd edn. Thomson-Brooks/Cole, Belmont (2007)

Tukey, J.W.: Exploratory Data Analysis. Addison-Wesley Publishing Co., Reading (1977)

Walpole, R.E., Myers, R.H., Myers, S.L., Ye, K.: Probability and Statistics for Engineers and Scientists, 9th edn. Pearson Education, Essex (2016)

Chapter 3
Linear Regression and Correlation

3.1 Simple Linear Regression Model

Many problems in engineering, agriculture, medicine, economy and social sciences involve exploring the inherent relationships between two or more variables. For example, in canola cultivation, it may be known that the crop yield response is related to the nitrogen fertilizer quantities. It may be of interest to develop a method of prediction, that is, a procedure for estimating the canola crop yield response for various levels of nitrogen fertilizer. A crop yield response is an example of *dependent* or *response variable*, nitrogen fertilizer quantity is an *independent or explanatory variable*. Depending on the context, an independent variable is sometimes called a predictor or input variable, regressor, covariate, control variable (in econometrics) or feature (in machine learning). Except for response variable there are also other synonyms in statistics for dependent variable: regressand, explained, output/outcome or predicted variable. A reasonable form of a relationship between the crop yield response Y and the nitrogen fertilizer quantity x, is the linear relationship:

$$Y = a_1 x + a_0 \qquad (3.1.1)$$

where, a_0 is the *intercept* and a_1 is the *slope*.

If the relationship is exact, then it is a *deterministic relationship* between two variables and there is no random component to it. However, in the example above, as well as in countless other phenomenon, the relationship is not deterministic (i.e., a given x does not always give the same value for Y). Thus, the problem of modeling such relationships is *probabilistic* in nature since the relationship above cannot be viewed as being exact. In other words, there must be a *random component* to the equation that relates the variables. This random component takes into account considerations that are not being measured, or, are not understood by the scientists or

© Springer Nature Switzerland AG 2020
K. Stapor, *Introduction to Probabilistic and Statistical Methods with Examples in R*,
Intelligent Systems Reference Library 176,
https://doi.org/10.1007/978-3-030-45799-0_3

engineers. The linear equation above is an approximation, that is a simplification of something unknown and much more complicated. More often than not, the models that are simplifications of more complicated and unknown structures are linear in nature.

The concept of **regression analysis** deals with finding the best relationship between Y and x, quantifying the strength of that relationship, and, using methods that allow for prediction of response values given values of the explanatory variable. In many applications there will be more than one explanatory variable that helps to explain Y. The resulting analysis is termed as **multiple regression** while the analysis of the single regressor case is called **simple regression**.

The simple linear regression model assumes that the relationship between the measured variable x and random variable Y is as follows:

$$Y_i = a_1 x_i + a_0 + \varepsilon_i \quad i = 1, \ldots, n \tag{3.1.2}$$

where ε_i are random variables called *random errors*, n is the number of paired observations used to estimate the model. In addition, random errors are assumed to meet the following conditions:

$$E(\varepsilon_i) = 0 \tag{3.1.3}$$

$$D^2(\varepsilon_i) = \sigma^2 \tag{3.1.4}$$

$$\text{cov}(\varepsilon_i, \varepsilon_j) = 0 \quad i \neq j \tag{3.1.5}$$

The symbols a_0, a_1, σ are the unknown *parameters* of the model. The first two are intercept and line slope, the third, when squared is called error variance. The simple linear regression model described in the above formula consists of two parts:

- A *systematic component* $a_1 x_i + a_0$ expressing the effect of x on variable Y,
- A *random component* ε_i expressing the combined effect of all other variables or factors, except the variable x affecting variable Y.

From the assumptions above, several things become visible. The first condition implies that at a specific x, the y values are distributed around the true or *population regression line* $E(Y|X = x) = a_1 x + a_0$, i.e. the true regression line goes through the means of the responses, and actual observations are on the distribution around the means. The second condition is often called a *homogeneous variance assumption,* and the last means there is no correlation between the individual random errors. For the purpose of inferential procedures, we shall need to add one assumption:

$$\varepsilon_i \sim N(x; 0, \sigma^2) \tag{3.1.6}$$

In the simple linear regression model, the values of the independent variable x are treated as pre-determined or non-random values. In the above-mentioned example of canola cultivation, we perform the procedure of seeding the n experimental plots with the pre-determined quantities $x_1, \ldots, x_n$ of nitrogen fertilizer. After some time, we measure the crop yields responses $y_1, \ldots, y_n$ from the seeded experimental plots. The experimental (regression) data comprise the n-element sample of paired observations $\{(x_1, y_1), \ldots, (x_n, y_n)\}$ that can be plotted in a *scatter diagram* (see for example Fig. 3.2) indicating visually, if our assumption of linearity between regressor x and the response Y appears to be reasonable.

It should be noted that the model described above is *conceptual* in nature: as we never observe the actual values of random errors in practice, we can never draw the *true regression line* $E(Y|X = x) = a_0 + a_1 x$, only the estimated line can be drawn.

An important aspect of regression analysis is to estimate the parameters of the model, based on n-element sample which will be presented in the next section.

3.1.1 Estimation of the Parameters

The *method of least squares* is used to estimate the parameters of simple linear regression model. Before we present this method, we should introduce the concept of a residual.

Given a set of regression data $\{(x_i, y_i) : i = 1, \ldots, n\}$ and already fitted (estimated) regression model:

$$\hat{y} = \hat{\alpha}_0 + \hat{\alpha}_1 x \qquad (3.1.1.1)$$

(see Fig. 3.1), the *i*th *residual* is the difference between the *i*th observed response value and the ith response value that is predicted from linear regression model which

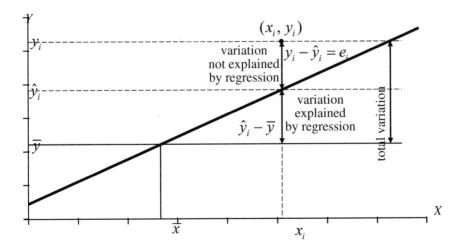

Fig. 3.1 The variations for an observation in the linear regression model

can be written as:

$$e_i = y_i - \hat{y}_i \quad i = 1, \ldots, n \tag{3.1.1.2}$$

Obviously, small residuals are a sign of a good fit. It should be noted that the ε_i are not observed but e_i are observed and play an important role in the regression analysis. The sum of squares of regression residuals for all observations in the sample is denoted by SSE (*Sum of Square Errors*, sometimes *Residual Sum of Squares*):

$$SSE = \sum_{i=1}^{n} e_i^2 = \sum_{i=1}^{n} (y_i - \hat{y}_i)^2 \tag{3.1.1.3}$$

The *method of least squares* finds the estimators of parameters α_0 and α_1, so that the sum of squares of the residuals SSE is a minimum. Hence, we shall find α_0 and α_1, so as to minimize

$$SSE = \sum_{i=1}^{n} e_i^2 = \sum_{i=1}^{n} (y_i - a_1 x_i - a_0)^2 \tag{3.1.1.4}$$

Differentiating *SSE* with respect to α_0 and α_1, and setting the partial derivatives to zero, we have:

$$\frac{\partial SSE}{\partial a_0} = 2 \sum_{i=1}^{n} (y_i - a_1 x_i - a_0)(-1) = 0 \tag{3.1.1.5}$$

$$\frac{\partial SSE}{\partial a_1} = 2 \sum_{i=1}^{n} (y_i - a_1 x_i - a_1)(-x_i) = 0 \tag{3.1.1.6}$$

Rearranging the terms, we obtain the so-called *normal equations*:

$$a_1 \sum_{i=1}^{n} x_i + a_0 n = \sum_{i=1}^{n} y_i \tag{3.1.1.7}$$

$$a_1 \sum_{i=1}^{n} x_i^2 + a_0 \sum_{i=1}^{n} x_i = \sum_{i=1}^{n} x_i y_i \tag{3.1.1.8}$$

The solution of this linear system gives the following estimators $\hat{a}_1$, $\hat{a}_0$ of the parameters a_0, a_1:

$$\hat{a}_1 = \frac{n \sum_{i=1}^{n} x_i y_i - \sum_{i=1}^{n} x_i \sum_{i=1}^{n} y_i}{n \sum_{i=1}^{n} x_i^2 - \left(\sum_{i=1}^{n} x_i\right)^2} = \frac{\sum_{i=1}^{n} (x_i - \bar{x})(y_i - \bar{y})}{\sum_{i=1}^{n} (x_i - \bar{x})^2} \tag{3.1.1.9}$$

$$\hat{a}_0 = \bar{y} - \hat{a}_1 \bar{x} \tag{3.1.1.10}$$

where:

$$\bar{x} = \frac{1}{n}\sum_{i=1}^{n} x_i \quad \bar{y} = \frac{1}{n}\sum_{i=1}^{n} y_i \tag{3.1.1.11}$$

Introducing the following denotations:

$$SS_x = \sum_{i=1}^{n} (x_i - \bar{x})^2 = \sum_{i=1}^{n} x_i^2 - \frac{\left(\sum_{i=1}^{n} x_i\right)^2}{n} \tag{3.1.1.12}$$

$$SS_y = \sum_{i=1}^{n} (y_i - \bar{y})^2 = \sum_{i=1}^{n} y_i^2 - \frac{\left(\sum_{i=1}^{n} y_i\right)^2}{n} \tag{3.1.1.13}$$

$$SS_{xy} = \sum_{i=1}^{n} (x_i - \bar{x})(y_i - \bar{y}) = \sum_{i=1}^{n} x_i y_i - \frac{\left(\sum_{i=1}^{n} x_i\right)\left(\sum_{i=1}^{n} y_i\right)}{n} \tag{3.1.1.14}$$

we can rewrite the derived formulas for model parameters in the following form: slope of the regression line:

$$\hat{a}_1 = \frac{SS_{xy}}{SS_x} \tag{3.1.1.15}$$

intercept:

$$\hat{a}_0 = \bar{y} - \hat{a}_1 \bar{x} \tag{3.1.1.16}$$

It can be shown that these least squares estimates, $\hat{a}_0$, $\hat{a}_1$, are unbiased.

The third, yet not estimated parameter of the simple linear regression model is the error or model variance σ^2 which measures squared deviations between Y values and their mean (i.e. true regression line $E(Y|X = x) = a_1 x + a_0$). The unbiased estimator of the parameter σ^2 is a *residual variance* (sometimes called a *Mean Square Error, MSE*), defined as follows:

$$\hat{\sigma}^2 = MSE = \frac{SSE}{n - 2} \tag{3.1.1.17}$$

The denominator $n - 2$ is the quantity called the *degrees of freedom* associated with the estimator (two parameters are estimated).

The fitted regression model allows for the computation of predicted values from the fitted line $\hat{y} = \hat{a}_0 + \hat{a}_1 x$. Other types of analyses and diagnostic information that will ascertain the strength of the relationship and the adequacy of the fitted model are needed. This problem will be discussed later. Also, checking if the linear model assumptions stated at the beginning of the chapter are met is necessary (here, the plot function in R environment can be helpful).

Inferences Concerning Model Parameters
For the purposes of prediction, one may also be interested in drawing certain inferences about the estimators of slope and intercept. It can be shown that the estimates of standard errors of the estimators $\hat{a}_0$ and $\hat{a}_1$ are

$$se(\hat{a}_0) = \frac{RSE\sqrt{\sum_{i=1}^n x_i^2}}{\sqrt{n \cdot SS_x}} \qquad (3.1.1.18)$$

$$se(\hat{a}_1) = \frac{RSE}{\sqrt{SS_x}} \qquad (3.1.1.19)$$

where the standard error of model variance, so called residual standard error is:

$$RSE = \sqrt{MSE} \qquad (3.1.1.20)$$

Standard errors can be the used to construct confidence intervals on the confidence level α for the regression parameters:

$$\hat{a}_0 \pm t_{\alpha/2,n-2} se(\hat{a}_0) \qquad (3.1.1.21)$$

$$\hat{a}_1 \pm t_{\alpha/2,n-2} se(\hat{a}_1) \qquad (3.1.1.22)$$

Of special importance is testing the null hypothesis about a slope of regression line:

$$H_0: \quad a_1 = 0 \qquad (3.1.1.23)$$

versus

$$H_1: \quad a_1 \neq 0 \qquad (3.1.1.24)$$

The following test statistic which follows the t-distribution with $(n-2)$ degrees of freedom is used in this test:

$$t = \frac{\hat{a}_1}{s(\hat{a}_1)} \qquad (3.1.1.25)$$

When the null hypothesis is not rejected, the conclusion is that there is no significant linear relationship between Y and the independent variable x. Rejection of H_0 indicates that the linear term in x residing in the model, explains a significant portion of variability in Y.

3.1.2 Analysis of Variance and the Accuracy of Fit

Assessing the quality of the estimated regression line is often performed by *analysis-of-variance* (ANOVA) *approach*. This relies on the partitioning of the total sum of squares of y (i.e. total variation of y) designated here as SST into the two components (see Fig. 3.1):

$$\sum_{i=1}^{n} (y_i - \bar{y})^2 = \sum_{i=1}^{n} (\hat{y}_i - \bar{y})^2 + \sum_{i=1}^{n} (y_i - \hat{y}_i)^2$$
$$\text{SST} = \quad \text{SSR} + \quad \text{SSE} \tag{3.1.2.1}$$

where:

$$SST = \sum_{i=1}^{n} (y_i - \bar{y})^2 \tag{3.1.2.2}$$

$$SSR = \sum_{i=1}^{n} (\hat{y}_i - \bar{y})^2 \tag{3.1.2.3}$$

$$SSE = \sum_{i=1}^{n} (y_i - \hat{y}_i)^2 \tag{3.1.2.4}$$

The first component, SSR, is called the *regression sum of squares* and it reflects the amount of variation in the y values explained by the straight, line. The second component is the already introduced *Sum of Square Errors* SSE, which reflects the variation about the regression line (i.e. the part of the total variation not been explained by regression).

The analysis of variance is conveniently presented in the form of the following Table 3.1.

Suppose now that we are interested in testing the following null hypothesis:

$$H_0 : a_1 = 0 \tag{3.1.2.5}$$

that the variation in Y results from chance—it is independent of the values of x, against the alternative:

Table 3.1 The table for the analysis of variance in regression

Source of variation	Sum of squares	Degrees of freedom	Mean squares	F statistic
Regression	SSR	k	$MSR = \frac{SSR}{k}$	$F = \frac{MSR}{MSE}$
Error	SSE	$n - k - 1$	$MSE = \frac{SSE}{n-k-1}$	
Sum	SST	$n - 1$	$MST = \frac{SST}{n-1}$	

$$H_1 : a_1 \neq 0 \qquad (3.1.2.6)$$

The value of the F statistic (a test statistic here):

$$F = \frac{MSR}{MSE} \qquad (3.1.2.7)$$

is calculated based on the analysis of variance table (the last column of this table). The critical region for this test statistic is:

$$\langle F(1 - \alpha, 1, n - 2), +\infty) \qquad (3.1.2.8)$$

When the null hypothesis is rejected, that is, when the computed F-statistic exceeds the critical value $F(1 - \alpha, 1, n - 2)$, we conclude that there is a significant amount of variation in the response resulting from the postulated straight-line relationship. The linear relationship between the variable Y and x is statistically significant at the significance level α.

If null hypothesis is not rejected, we conclude that the data did not reflect sufficient evidence to support the model postulated. In case of simple regression, the above test is equivalent to the previously presented t-test.

The already introduced RSE is an absolute measure of lack of fit of linear model to the data. The alternative measure is the R^2 statistic, also called *the coefficient of determination*, which is a measure of goodness of fit of a model to the observed data:

$$R^2 = \frac{SSR}{SST} = 1 - \frac{SSE}{SST} \qquad (3.1.2.9)$$

The R^2 statistic (pronounced "R-squared") is a measure of the proportion of variability explained by the fitted model. The values of this coefficient are in the range [0,1]. The higher the value, the better the fit: if the fit is perfect, all residuals are zero, and thus $R^2 = 1$.

When comparing the competing models based on R^2 for the same data set, the R^2 can be made artificially high by adding additional regressors (it decreases SSE and thus increases R^2), thus, using it in multiple regression—an adjusted version should be applied.

3.1.3 Prediction

One reason for building a linear regression model is to predict response values at one or more values x_0 of the independent variable x. The equation $\hat{y} = \hat{\alpha}_0 + \hat{\alpha}_1 x$ may be used to predict the mean response $E(Y|x_0)$ (i.e. a conditional expectation $E(Y|X = x)$) at a prechosen value $x = x_0$, or it may be used to predict a single value y_0 of the variable Y_0, when $x = x_0$. As one would expect, the prediction error to be higher in the case of a single predicted value y_0 than in the case where a mean $E(Y|x_0)$ is predicted. This, of course, will affect the width of intervals for the predicted values.

In this section, the focus is on errors associated with the prediction.

A $100 \cdot (1 - \alpha)\%$ *confidence interval* for the mean response $E(Y|x_0)$ is:

$$\hat{y}_0 \pm t(1 - \alpha/2, n - 2) \cdot RSE \cdot \sqrt{\frac{1}{n} + \frac{(x - \bar{x})^2}{SS_x}} \qquad (3.1.3.1)$$

where $t(1 - \alpha/2, n - 2)$ is a quantile of t-distribution with $(n - 2)$ degrees of freedom.

The interpretation of the above formula is as follows. Recall the experiment of canola cultivation from the beginning of the chapter. Suppose we have decided to start seeding the canola plots on a larger scale, and repeat many times the seeding at the fertilizer quantity $x = x_0$. Then, using the above formula, the average amount of crop yield together with the accuracy can be calculated.

Another type of interval that is often confused with that given above for the mean, is the prediction interval for a future observed response y_0.

A $100 \cdot (1 - \alpha)\%$ *prediction interval* for a single response y_0 is:

$$\hat{y}_0 \pm t(1 - \alpha/2, n - 2) \cdot RSE \cdot \sqrt{1 + \frac{1}{n} + \frac{(x_0 - \bar{x})^2}{SS_x}} \qquad (3.1.3.2)$$

There is a distinction between the concept of a confidence interval and the prediction interval described above. The *confidence interval* interpretation is identical to that described for all confidence intervals on population parameters discussed in the previous chapter. This follows from the fact that $E(Y|x_0)$ is a population parameter. The *prediction interval*, however, represents an interval that has a probability equal to $(1 - \alpha)$ of containing not a parameter, but a future value y_0 of the random variable Y.

Prediction intervals are always wider than confidence ones because they incorporate both the error in the estimate of the model function (*reducible error*) and the uncertainty as to how much an individual point will differ from the population model, here, the regression line (*irreducible error*).

Example 3.1.3.1 Suppose we wish to examine the relationship between the two variables: the canola crop yield (dependent variable Y) and the quantity of nitrogen fertilizer used (independent variable x). The $n = 12$ experimental plots with different levels of nitrogen fertilizer were seeded. The obtained experimental data are given in Table 3.2.

We assume linear relationship between the canola crop yield and the quantity of nitrogen fertilizer which follows from the scatter diagram (Fig. 3.2). Using the introduced in 3.1.1 formulas for the parameter estimators, we obtain the following estimates of model parameters (on the significance level $\alpha = 0.05$):

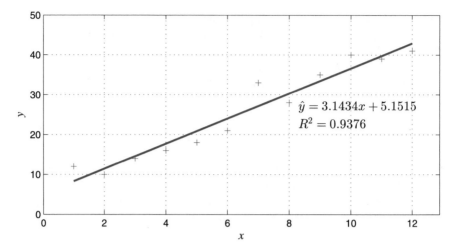

Fig. 3.2 Scatter diagram for the canola crop yield data with the regression line superimposed

$$\hat{\alpha}_1 = 3.146 \pm 0.256$$

$$\hat{\alpha}_0 = 5.152 \pm 0.545$$

$$MSE = 9.398 \pm 3.066$$

The estimated regression line is:

$$\hat{y} = 3.143x + 5.152$$

The calculated values of variations (i.e. sums of squares), degrees of freedom, and average squares are included in the analysis of variance table (Table 3.3). The value of the test statistic in the F test is:

$$F = \frac{SSR}{MSE} = 150.345 > F(1 - 0.05, 1, 10) = 4.96$$

And, as we see, it falls into the critical region. Hence, at the significance level $\alpha = 0.05$, we reject the null hypothesis that the slope of the regression line is equal to zero. Thus, the examined linear relationship between the canola crop yield and fertilizer's quantities is statistically significant.

The coefficient of determination is:

$$R^2 = \frac{SSR}{SST} = 0,938$$

Table 3.2 Experimental data: fertilizer level x, canola crop yield y

x	y
1	12
2	10
3	14
4	16
5	18
6	21
7	33
8	28
9	35
10	40
11	39
12	41

Table 3.3 The analysis of variance table for canola regression

Source of variation	Sum of squares	Degrees of freedom	Mean squares	F statistic
SSR	1412, 939	1	1412, 939	$F = 150, 345$
SSE	93, 978	10	9, 398	
SST	1506, 917	11	136, 992	

Almost 94% of the canola crop yield variation can be predicted from fertilizer's quantity. It follows that our linear regression model fits well to the experimental data.

Calculating the confidence interval for each experimental point x_i, we obtain two series of points resulting in *confidence limits* for the mean $E(Y|X = x)$, which is shown in Fig. 3.3.

3.2 Multiple Linear Regression

In most real-life problems, the complexity of their mechanisms is such that in order to be able to predict an important response, more than one independent variable is needed. If the regression analysis is to be applied, this means that a *multiple regression model* is needed. When this model is linear in the coefficients, it is called a *multiple linear regression model*. For the case of k independent variables (regressors) $x_1, \ldots, x_k$, the mean $Y|x_1, \ldots, x_k$ is given by the *multiple linear regression model*:

$$E(Y|x_1, \ldots, x_k) = a_0 + a_1 x_1 + \ldots + a_k x_k \tag{3.2.1}$$

and the estimated response is obtained from the sample regression equation:

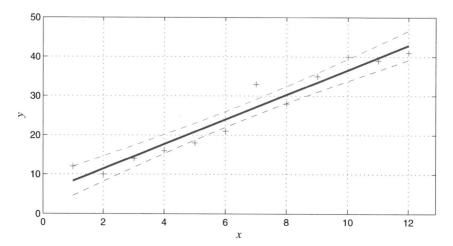

Fig. 3.3 Confidence limits (dashed line) for E(Y|x) in canola crop yield data

$$\hat{y} = \hat{a}_0 + \hat{a}_1 x_1 + \ldots + \hat{a}_k x_k \qquad (3.2.2)$$

Each regression coefficient a_i is estimated by $\hat{a}_i$ from the sample data:

$$\{(x_{1i}, x_{2i} \ldots, x_{ki}, y_i); i = 1, \ldots, n \, n > k\} \qquad (3.2.3)$$

where y_i is the observed response to the values $x_{1i}, x_{2i} \ldots, x_{ki}$ of the k independent variables $x_1, x_2 \ldots, x_k$. Similar to simple regression, it is assumed that each observation in a sample satisfies the following condition:

$$y_i = a_0 + a_1 x_1 + \ldots + a_k x_k + \varepsilon_i \qquad (3.2.4)$$

where ε_i are the random errors, independent and identically distributed with mean zero and common variance σ^2.

Using *the method of least squares* to obtain the estimates $\hat{a}_i$, we minimize the expression:

$$SSE = \sum_{i=1}^{n} e_i^2 = \sum_{i=1}^{n} (y_i - \hat{y}_i)^2 \qquad (3.2.5)$$

Differentiating SSE in turn with respect to $a_i \, i = 1, \ldots, k$ and equating to zero, we generate the set of $k + 1$ normal estimation equations for multiple linear regression. These equations can be solved by any appropriate method for solving systems of linear equations. Statistical packages are often used here to obtain numerical solutions of these equations.

The mathematical manipulations in fitting a multiple linear regression model can be considerably facilitated by using notation of matrix theory. The above introduced model with k regressors and n equations describing how the response is generated, can be written using matrix notation as:

$$\begin{pmatrix} Y_1 \\ \vdots \\ Y_n \end{pmatrix} = \begin{pmatrix} 1 & x_{11} & \cdots & x_{1k} \\ \vdots & \vdots & \vdots & \vdots \\ 1 & x_{n1} & \cdots & x_{nk} \end{pmatrix} \cdot \begin{pmatrix} a_0 \\ \vdots \\ a_k \end{pmatrix} + \begin{pmatrix} \varepsilon_1 \\ \vdots \\ \varepsilon_n \end{pmatrix}$$
$$Y \quad = \quad Xa \quad + \quad \varepsilon \tag{3.2.6}$$

where the symbol Y is the column vector of the response variables and a is a vector of unknown model coefficients.

Obtaining an unambiguous solution requires that the matrix $X^T X$ be non-singular, hence the requirement that the row $r(X)$ of the matrix X satisfy the following condition: $r(X) = k + 1 \leq n$. Then, it can be shown that the least square estimate of vector a of coefficients is given by:

$$\hat{a} = (X^T X)^{-1} X^T Y \tag{3.2.7}$$

The least squares estimator $\hat{a}$ possesses a minimum variance property among all unbiased linear estimators (this follows from *the Gauss-Markov theorem*).

Making the additional assumption that $\varepsilon \sim N(0, \sigma^2 I)$, by using the maximum likelihood method, we can obtain the following estimator of unknown variance of errors:

$$\hat{\sigma}^2 = \frac{1}{n}(y - X\hat{a})^T (y - X\hat{a}) \tag{3.2.8}$$

The maximum likelihood estimate of *vector a* is the same.

Similarly, as in simple the linear regression, in multiple linear regression model the total variability of Y can be decomposed into two components: variation explained by regression (SSR) and not explained (SSE). Based on this, the accuracy of the fit can be measured using a *coefficient of determination*:

$$R^2 = \frac{SSR}{SST} = 1 - \frac{SSE}{SST} \tag{3.2.9}$$

The quantity R^2 gives the proportion of the total variation in $y_i's$ "explained" by, or attributed to, the predictor variables $x_1, \ldots, x_k$.

The null hypothesis about the significance of the linear relationship between independent variables and the dependent one has the form:

$$H_0 : a_1 = \ldots = a_k = 0 \tag{3.2.10}$$

Table 3.4 The results of the experiment—measurement of the hardness y of the alloy for the fixed values of 2 components

$\times_1$	$\times_2$	y
30	12	52
24	9	36
28	18	80
55	14	84
46	22	95
28	16	74
44	8	62
56	14	86

The alternative hypothesis H_1 states that at least one of the coefficients $a_1, a_2, \ldots, a_k$ does not disappear. To verify that hypothesis, the following test statistic is used:

$$F = \frac{MSR}{MSE} = \frac{SSR/k}{SSE/(n-k-1)} \qquad (3.2.11)$$

The critical region of the above test statistic is:

$$\langle F(1-\alpha, k, n-k-1), \ +\infty) \qquad (3.2.12)$$

Example 3.2.1 The influence of two different components X_1, X_2 on the hardness Y of a certain alloy was investigated. The results for 8 measurements with pre-determined values of variables X_1, X_2 (the amounts of the components) are shown in the Table 3.4 below. Assuming a linear dependency between the hardness of alloy and the amounts of their two components, based on the sample data from that table, we estimate the coefficients of the multiple linear regression model at the significance level $\alpha = 0.05$.

Unknown coefficients a_0, a_1, a_2 of the model are calculated according to the formula:

$$\hat{a} = (X^T X)^{-1} X^T y$$

Using the calculated values, we obtain the following equation describing the relationship between amounts of alloy components and its hardness:

$$\hat{y} = -5,029 + 0,846 x_1 + 3,062 x_2$$

The analysis of variance table presents the two sources of variation in alloy hardness: that explained by the regression (SSR) and SSE. (Table 3.5).

The quantile of the F distribution at the significance level $\alpha = 0.05$ is $F(1 - 0.05, 2, 6) = 5,786$ which gives the critical region:

Table 3.5 The analysis of variance table for the alloy hardness data

Source of variation	Sum of squares	Degrees of freedom	Mean squares	F statistics
SSR	2507.155	2	MSR = 1253.577	F = 28,527
SSE	219.720	5	MSE = 43.944	
SST	2726.875	7	MST = 389.554	

$$(5.786, +\infty)$$

The value of the F statistic falls into the critical region, thus the null hypothesis, is rejected at the significance level $\alpha = 0.05$. The linear relationship between alloy hardness and the amounts of its two components is, thus, statistically significant. The coefficient of determination is:

$$R^2 = \frac{SSR}{SST} = 0.92$$

Multiple linear regression of two predictors explains 92% of the variability in alloy hardness. It follows that our linear model fits well to the experimental data.

3.3 Correlation

So far in this chapter we have assumed that the independent regressor variable x is a physical variable but not a random variable. In many applications of regression techniques it is more realistic to assume that both X and Y are random variables, and the measurements $\{(x_i, y_i) : i = 1, 2, \ldots, n\}$ are observations from a population having the joint density function $f(x, y)$. We shall consider the problem of measuring the relationship between the two variables X and Y. For example, if X represents worker's salary and Y—his efficiency, then we may suspect that larger values of X are associated with larger values of Y and vice versa. This is an example of positive *correlation*, which is a relationship between the two variables in which both variables move in tandem—that is, in the same direction. If X represents one's time of working and Y—one's free time, then the more time one works the less free time one has. This is a *negative correlation*, which is a relationship between two variables whereby they move in opposite directions. The already introduced scatter plots are the visualization of the correlation. Correlation is said to be linear if the ratio of change of the two variables is constant (for ex. when worker's salary is doubled by doubling his efficiency). When the ratio of change is not constant, we say about non linear correlation.

In statistics, *correlation* is any statistical association, or, dependence between the two random variables (whether *causal* or not), though it commonly refers to the degree to which a pair of variables are linearly related. Formally, two random variables are *dependent* if they do not satisfy a mathematical property of probabilistic independence.

There are several correlation coefficients, often denoted by ρ(*rho*), measuring the degree of correlation. The most common of these is the already introduced in chapter one *Pearson correlation coefficient*. It is sensitive only to a linear relationship between two variables, and one should bear in mind that it may be present even when one variable is a nonlinear function of the other.

Population Pearson correlation coefficient has been defined as:

$$\rho = \frac{\text{cov}(X, Y)}{D(Y) \cdot D(Y)} \tag{3.3.1}$$

where $D(X)$, $D(Y)$ are standard deviations of X and Y, respectively, while $\text{cov}(X, Y)$ is a covariance of X and Y, already defined.

Pearson correlation coefficient, when applied to a sample, may be referred to as the *sample Pearson correlation coefficient* (often denoted by r). We can obtain a formula for it by substituting estimates of the covariances and variances into the formula given above.

Given a sample, i.e. the paired data $\{(x_i, y_i) : i = 1, 2, \ldots, n\}$, a *sample Pearson's correlation coefficient* is defined as:

$$r = \frac{\sum_{i=1}^{n} (x_i - \bar{x})(y_i - \bar{y})}{\sqrt{\sum_{i=1}^{n} (x_i - \bar{x})^2} \cdot \sqrt{\sum_{i=1}^{n} (y_i - \bar{y})^2}} = \frac{SS_{xy}}{\sqrt{SS_x SS_y}} \tag{3.3.2}$$

The Pearson correlation coefficient ranges from -1 to 1. A value of 1 implies that a linear equation describes the relationship between X and Y perfectly, with all data points lying on a line for which Y increases as X increases. A value of -1 implies that all data points lie on a line for which Y decreases as X increases. A value of 0 implies that there is no linear correlation between the variables.

It might be said, then, that sample estimates of ρ close to unity in magnitude imply good correlation, or, linear association between X and Y, whereas values near zero indicate little or no correlation. But the interpretation of a correlation coefficient depends on the context and purposes. A correlation of 0.8 may be very low if one is verifying a physical law using high-quality instruments, but may be regarded as very high in the social sciences, where there may be a greater contribution from complicating factors.

The existence of a linear relationship between two variables in a population can be investigated using methods of statistical inference: testing hypothesis about population correlation coefficient ρ, or, constructing a confidence interval that, on repeated sampling, has a given probability of containing ρ.

We present only the first type of inference, a t-test, in which one aims to test the null hypothesis.

$$H_o : \rho = 0 \tag{3.3.3}$$

that the true (population) correlation coefficient ρ is equal to 0 versus an appropriate alternative. The test statistic:

$$t = \frac{r\sqrt{n-2}}{\sqrt{1-r^2}} \qquad (3.3.4)$$

(which is based on a value of a sample correlation coefficient r), has a t distribution with $n - 2$ degrees of freedom. A critical region is constructed similarly to other t-tests.

This test is equivalent to testing hypothesis $H_o : a_1 = 0$ for the simple linear regression model.

It is important to remember that the correlation coefficient between two variables is a measure of their linear relationship, and, that a value of $r = 0$ implies a lack of linearity and not a lack of association.

Bibliography

Draper, N.R., Smith, H.: Applied Regression Analysis, 3rd edn. Wiley, New York (1998)
Rao, C.R.: Linear Statistical Inference and its Applications, 2nd edn. Wiley, New York (1973)

Appendix A
Permutations, Variations, Combinations

In order to calculate a probability of an event, we shall "count" the number of sample points, i.e. simple events of sample space that favour a given event. In many cases, as the number of such favouring points is very large, we should be able to solve this problem without actually listing each favouring element (i.e. *counting without counting*). These problems belong to a branch of mathematics, called *combinatorics* which is concerned with the problems like how to determine the number of possible configurations (arrangements) of objects of a given type. Here, we describe three such configurations: permutations, combinations and variations.

Permutation (*without repetition*) is an arrangement of all objects, where order does matter. The number of permutations of n objects is:

$$P_n = n! \tag{A.1}$$

When some of those objects are identical copies, this is a *permutation with repetition*. The number of such permutations of n objects of which n_1 are of the 1st kind, n_2 are of the 2nd kind, ..., n_k are of the k-th kind is:

$$P_n^{n_1, n_2, \ldots, n_k} = \frac{n!}{n_1! \cdot n_2! \cdot \ldots n_k!} \tag{A.2}$$

In many problems, we are interested in the number of ways of selecting k objects from n without regard to order. These selections are called *combinations*. The number of k-element combinations of n objects without repetitions is:

$$C_n^k = \binom{n}{k} = \frac{n!}{k!(n-k)!} \tag{A.3}$$

and with repetitions:

© Springer Nature Switzerland AG 2020
K. Stapor, *Introduction to Probabilistic and Statistical Methods with Examples in R*,
Intelligent Systems Reference Library 176,
https://doi.org/10.1007/978-3-030-45799-0

$$\overline{C}_n^k = \binom{n+k-1}{k} \tag{A.4}$$

Sometimes, we may choose k elements in a specific order from n element set. Each such choice is called a k-element *variation*. To count k-element variations of n objects, we first need to choose a k-element combination, and then a permutation of the selected objects. The number of k-element variations of n objects is, thus:

$$V_n^k = \binom{n}{k} \cdot k! = \frac{n!}{(n-k)!} \tag{A.5}$$

If repetitions are allowed, the counting procedure is as follows. The 1st object can be selected in n ways and is then returned, then the same is true for the 2nd object and so on until the k-th object. This gives the following number of k-element variations of n objects (with repetitions):

$$\overline{V}_n^k = n^k \tag{A.6}$$

Appendix B
An Introduction to R

About R

R is a scripting language for statistical data manipulation and analysis. It was inspired by, and is mostly compatible with, the statistical language S developed by AT&T (S obviously standing for statistics). R is sometimes called "GNU S," to reflect its open source nature. R is available for Windows, Macs, Linux. R also provides an environment in which you can produce graphics. As a programming language, R has a large libraries of pre-defined functions that can be used to perform various tasks. A major focus of these pre-defined functions is statistical data analysis, and these allow R to be used purely as a toolbox for standard statistical techniques. However, some knowledge of R programming is essential to use it well at any level. Therefore, in this Appendix, we learn about common data structures and programming features in R. For more resources, see the R Project homepage:

http://www.r-project.org,

which links to various manuals and other user-contributed documentation.

One typically submits commands to R via text in a terminal window (a console), rather than mouse clicks in a Graphical User Interface. If you can't live without GUIs, you use one of the free GUIs that have been developed for R, e.g. *R Studio*.

Installation, libraries

To install R on your computer, go to the home website of R:

http://cran.r-project.org

and do the following (assuming you work on a windows computer): (1) click download CRAN in the left bar, (2) choose a download site, then choose Windows as target operation system and click base, (3) choose *Download R x.x.x for Windows* (x.x.x stands for an actual version nr) and choose default answers for all questions.

R can do many statistical and data analyses. They are organized in the so-called *packages* or *libraries*. With the standard installation, most common packages are installed. To get a list of all installed packages, go to the packages window or type

© Springer Nature Switzerland AG 2020   153
K. Stapor, *Introduction to Probabilistic and Statistical Methods with Examples in R*,
Intelligent Systems Reference Library 176,
https://doi.org/10.1007/978-3-030-45799-0

library() in a console window. There are many more packages available on the R website. If you want to install and use a package, for example, the package called "*utils*", you should first install the package:

```
> install.packages(utils)
```

The, you should load the package by typing:

```
> library(utils)
```

Some simple examples

Your *working directory* is the folder on your computer in which you are currently working. Before you start working, please set your working directory to where all your data and script files are using *setwd(.)* command, for example type in the command window:

```
> setwd("C:/R/")
```

R can be used as a calculator. You can just type your equation in the command window, just after the ">":

```
> 1+1
```

and R will give the answer:

```
[1] 2
```

You can also give numbers a name. By doing so, they become so-called variables which can be used later. For example, you can type in the command window:

```
> a = 2
```

The variable *a* appears now in the *workspace*, which means that R now remembers what *a* is 2. You can also ask R what *a* is:

```
> a
[1] 2
```

To remove all variables from R's memory, type:

```
> rm(list=ls())
```

There is a large amount of documentation and help available. Some help is automatically installed. Typing in the console window the command:

```
> help(command-name)
```

gives help on the *command-name* function. It gives a description of the function, possible arguments and the values that are used as default for optional arguments.

R organizes numbers in s*calars* (a single number—0-dimensional), *vectors* (a row of numbers—1-dimensional) and *matrices* (like a table—2-dimensional). The *a* you have defined before is a scalar. To define a *vector* with the numbers 1, 2, 3 you need the function *c*, which is short for *concatenate*:

```
>b=c(1,2,3)
```

Matrices are nothing more than 2-dimensional vectors. To define a matrix, use the function *matrix*:

```
> m = matrix(data = c(1,2,3,4,5,6), ncol=3)
> m
```

```
     [, 1] [, 2] [, 3]
[1,]   1    2    3
[2,]   4    5    6
```

The argument *data* specifies which numbers should be in the matrix. Use either *ncol* to specify the number of columns or *nrow* to specify the number of rows.

A *data frame* is a matrix with names above the columns.

```
>f = data.frame(x=c(1,2,3), y=c(4.5.6))
```

You can for example select the column called *y* from the data frame called *f* for computing its mean using pre-defined function *mean*:

```
>mean(f$y)
```

The operator "$" allows you to extract elements by name from a named list.

There are many ways to write data from within the R environment to files, and to read data from files. The following commands first write the *f* data frame to a text file, called f1.txt and then read a file into a data frame *g*:

```
> write.table(f, file="f1.txt")
> g = read.table(file="f1.txt")
```

You do often automate your calculations using *functions*. Some functions are standard in R or in one of the packages. You can also program your own functions. Within the brackets you specify the *arguments*. As the example, the function *rnorm*, is a standard R function which creates random samples from a normal distribution. If you want 5 random numbers out of normal distribution with mean 1 and standard deviation 0.5 you can type:

```
> rnorm(5, mean=1, sd=0.5)
```

The R package also has built-in functions to calculate the values of density, and quantiles of the most commonly used distributions. Preceding the name of the distribution with the letter d, we obtain the value of density function of this distribution. For example, the following call:

```
> dnorm(0)
[1] 0.3989423
```

returns the value of probability density of normal distribution at point zero. The following function:

```
> qt(0.995,5)
[1] 4.032143
```

returns the quantile of t-distribution of order 0.995 with 5 degrees of freedom. R can make plots. The following lines show a simple plot:

```
> plot(rnorm(200), type="l", col="gold")
```

Two hundred random numbers are plotted by connecting the points by lines (type="l") in gold. Another example is the classical histogram plot, generated by the simple command:

```
> hist(rnorm(200))
```

which generates the histogram of 200 random numbers.

As in other programming languages, there are also programming statements for building a program in R. We present two examples. The *if*-statement is used when certain computations should only be done when a certain condition is met (and maybe something *else* should be done when the condition is not met).

```
> x=-1; y=0;
> if (x<0) {y=1} else {y=2}
```

In a *for-loop* you specify what has to be done and how many times. To tell "how many times", you specify a so-called *counter*, as in the following example:

```
> a=seq(from=1,to=4)
> b= c()
> for(i in 1:4) {b[i]=a[i]*2}
> b
2 4 6 8
```

Examples of statistical inference

Suppose we wish to verify a hypothesis $H_0 : m = 250$ about the mean weight of the products from a food factory: our random sample from a population is:

```
>dane = c(254, 269, 254, 248, 263, 256, 258, 261, 264, 258)
```

First, we need to verify if the sample comes from a normal population. The sample is small, therefore we use Shapiro-Wilk test. We install appropriate package first and then load it:

```
> install.packages("nortest")
> library(nortest)
```

Then we perform a test:

```
> shapiro.test(dane)
Shapiro-Wilk normality test
```

```
data: dane
W = 0.9853 , p-value = 0.9871
```

The very large p-value indicates that there is no evidence to reject null the hypothesis about normality. Then, we perform a t-test to verify a null hypothesis about the mean population weight:

```
> t.test(dane, mu=250)
One sample t-test
data: dane
t = 4.4764, df = 9, p-value = 0.00154
alternative hypothesis: true mean is not equal to 250
95 percentage confidence interval:
254.2045 262.7955
sample estimate of x
258.5
```

The low p-value (less than the test significance level $\alpha = 0.05$) indicates that the null hypothesis should be rejected. We also obtain a 95% confidence interval for the mean weight:

[254.2045 262.7955]

Different values of significance level can be set using $conf.level$ option.

Printed in the United States
by Baker & Taylor Publisher Services